Microsoft® Excel Manual for
Waner and Costenoble's

Applied Calculus
Second Edition

Larry J. Stephens
University of Nebraska at Omaha

Edwin C. Hackleman
Delta Software, Inc.

BROOKS/COLE

TM

THOMSON LEARNING

Australia • Canada • Mexico • Singapore • Spain • United Kingdom • United States

BROOKS/COLE

THOMSON LEARNING

Sponsoring Editor: *Curt Hinrichs*
Assistant Editor: *Seema Atwal*
Marketing Manager: *Karin Sandberg*
Editorial Assistant: *Nathan Day*
Production Coordinator: *Stephanie Andersen*

Permissions Editor: *Sue Ewing*
Cover Design: *Roy R. Neuhaus*
Cover Photo: *Walter Bibikow/FPG International*
Print Buyer: *Micky Lawler*
Printing and Binding: *Webcom Ltd.*

For more information about this or any other Brooks/Cole products, contact:
BROOKS/COLE
511 Forest Lodge Road
Pacific Grove, CA 93950 USA
www.brookscole.com
1-800-423-0563 (Thomson Learning Academic Resource Center)

For permission to use material from this work, contact us by
www.thomsonrights.com
fax: 1-800-730-2215
phone: 1-800-730-2214

Printed in Canada

10   9   8   7   6   5   4   3   2   1

ISBN: 0-534-38165-0

CONTENTS

PREFACE

We have prepared this book to augment the second edition of *Applied Calculus* by Stefan Waner and Steven R. Costenoble. Each chapter is divided into the major parts that correspond to the parent textbook, coupled with chapter review questions to tie everything together. Our approach is to employ a battery of tutorial exercises and to show precisely how you can solve them using Microsoft® Excel. In fact, solutions have been supplied for every problem, including the chapter review exercises. Numerous problems stemming from real-world applications have been prepared along with straightforward mathematical problems in an effort to capture and reflect the theme of the parent textbook. The table below provides a breakdown of the number of exercises and solutions written for each chapter:

Chapter	Tutorial	Chapter Review	Total
1	15	6	21
2	17	5	22
3	13	9	22
4	10	9	19
5	16	6	22
6	15	5	20
7	18	10	28
8	14	6	20
Total	118	56	174

The book has a host of special features:

- There are dozens of exercises that relate to a common body of information, such as a table of values or a scenario. The related problems are grouped immediately after the data are supplied, or they are written as one larger problem with several parts for you to answer. Structuring questions in this fashion effectively removes repetitive text. There are actually four to five times as many individual problems to solve than the table above implies.

- Variable names (x, y, z, p, α, β, μ, π, etc.) have been italicized to make them stand out better and to be as consistent as possible with the notation in the parent textbook.

- The spreadsheets and charts are exact screen captures of images that are cropped in an effort to zero in on the solution rather than show the entire window. This procedure saves space and eliminates noisy, wasted display matter.

We recommend that you follow a five-step procedure to maximize your learning efficiency:
1. Study at least one or two of the major divisions of each chapter of the parent textbook.
2. Work all of the tutorial problems stemming from the same divisions in this book while using Excel.
3. Repeat 1 and 2 until the textbook chapter and the tutorial problems is finished.
4. Solve as many of the chapter review exercises as you can to tie all the loose ends together. Do not look at the solution to any exercise until after you have made a serious effort to solve it.
5. Solve as many of the questions as you can in the parent textbook by using Excel, particularly those that the textbook authors recommend for spreadsheets or graphing calculators.

We recognize that your expertise in using Excel could range anywhere from beginner to power user. In either case, this book will still help you. Neophytes will learn anew, and veterans will be challenged to learn more. The earlier chapters and sections within a chapter tend to concentrate a bit more heavily on using Excel itself. However, the later chapters and sections assume that you have become familiar with the program interface, and the focus is then directed to a faster, more powerful solution to the problem.

Therefore, you might discover that the solutions to the problems in the earlier chapters and the sections within them are perhaps more detailed than they need be. However, as your skills improve within a chapter, many rudimentary steps are naturally skipped, and the problems themselves become more challenging. Ultimately, you may become so adept that your own solutions to the problems will be more efficient than the ones we have suggested. We have discovered from experience that the best tools may at first appear formidable. But once you master them, they become almost intuitive.

We have endeavored as a team to produce a book that will expand the horizon of using Excel, one of the most powerful computer programs ever written for mathematics. This software application has withstood the test of time and has grown more comprehensive with every new release. You will find that Excel will allow you to solve complex problems in minutes that would take hours to solve by hand or even by using a calculator. And, any shortcomings that we uncovered here using Excel 97 may have been overcome in the release you are now using.

We wish to thank Shirley J. Hackleman of Delta Software for her editorial assistance in preparing the hard copy manuscript.

LARRY J. STEPHENS
EDWIN C. HACKLEMAN

CHAPTER 1

Linear Functions and Models

1-1 Functions from the Numerical and Algebraic Viewpoints

Evaluating an Algebraically-Defined Function

1. Let f be the quadratic function defined by $f(x) = 2x^2 - 3x + 7$ with domain $(0, 500]$. Evaluate the function for values $x = 11, \ldots , 15$.

Solution: An Excel formula begins with an equal sign (=), followed by whatever the formula calculates. Excel always evaluates a formula from left to right, starting with the = sign. The result of the formula is then returned to a selected target cell. In this problem, all the values for x fall within the domain. Start by building a spreadsheet with the five data values, 11, 12, ... , 15 all in column A, rows 1 through 5:

	A
1	11
2	12
3	13
4	14
5	15

Enter the formula corresponding to the function in column B, row 1 (target cell B1):

	A	B
1	11	=2*A1^2-3*A1+7
2	12	
3	13	
4	14	
5	15	

When the formula, =2*A1^2-3*A1+7, is entered into the target cell B1, Excel calculates the value of the function for $x = 11$ and returns the result to cell B1:

	A	B
1	11	216
2	12	
3	13	
4	14	
5	15	

At this point, most of the work is done. You can easily apply the same formula to the rest of the values in column A. Simply select (highlight) cell B1 and position the mouse on the lower right-hand corner of the cell. A cross-hair cursor will appear. Now click, hold the mouse down, and drag the lower right-hand corner down through cells B2, B3, B4, and B5. The cells are highlighted as you drag. When you release the mouse, Excel will perform the required calculations and return the values to the target cells, B1:B5:

	A	B
1	11	216
2	12	259
3	13	306
4	14	357
5	15	412

As you dragged the corner down from row to row, Excel changed the subscripts within the formula to 2, 3, 4, and 5 automatically. To reveal this operation, click on any cell and examine the formula in the data entry textbox. Alternatively, the contents of cell B1 could have been copied and pasted into cell 2, then pasted into cell 3, and so on. However, dragging the corner of the cell with the mouse performs the same procedure quite a bit faster.

2. Evaluate the same quadratic function in problem 1 again by using the values of the function you obtained from the data.

Solution: The output from the function now becomes the input. Excel allows you to perform the calculations without entering the formula again. Simply select cell B1 and drag the corner to the right:

	A	B	C
1	11	216	92671
2	12	259	
3	13	306	
4	14	357	
5	15	412	

When you release the mouse, the value 92,671 will appear in cell C1. If you click on cell C1, the formula in the textbox shows the function again being applied to the value in cell B1. Now drag the lower right-hand corner of cell C1 down to cell C5:

	A	B	C
1	11	216	92671
2	12	259	133392
3	13	306	186361
4	14	357	253834
5	15	412	338259

As you drag the corner down, Excel calculates the same function with the values for B2 through B5. Release the mouse and the job is done. Click anywhere on the spreadsheet to remove the highlighting.

3. Let f be a higher-order function defined by $f(x) = 2x^3 - 3x^2 + \sqrt{5x} - 4$ with domain [0, 10]. Evaluate the function for values $x = 0, 1, 2, 3, 4$.

Solution: This function is more complicated than a quadratic but simple to handle using a spreadsheet program. All the data values for x fall within the domain. Start by building a spreadsheet with the five data values, 0, 1, ... , 4 all in column A, rows 1 through 5. Then enter the formula corresponding to the function in cell B1:

	A	B
1	0	=2*A1^3-3*A1^2+SQRT(5*A1)-4
2	1	
3	2	
4	3	
5	4	

When you enter the formula, =2*A1^3-3*A1^2+SQRT(5*A1)-4, Excel will calculate the value of the function and display the result, –4, in cell B1. Now click on cell B1, hold the mouse down, and drag the lower right-hand corner down to cell B5:

	A	B
1	0	-4
2	1	-2.763932023
3	2	3.16227766
4	3	26.87298335
5	4	80.47213595

Release the mouse, and the job is done. Cells B1 through B5 contain the values of the function for $x = 0$, 1, 2, 3, and 4.

Note: When Excel evaluates a formula containing more than one operator, it performs the operations in predefined order, starting from top to bottom, as listed in the following table:

Operator	Description
–	Negation (as in –1)
%	Percent
^	Exponentiation
* and /	Multiplication and Division
+ and –	Addition and Subtraction
&	Text Concatenation
= <> <= >= <>	Comparison

Therefore, you will likely encounter occasions when the formula calculations do not produce results that you would first expect according to the rules of algebra. A classic case is the evaluation of a formula that starts with a minus sign. For example, suppose you wish to evaluate the quadratic function defined by $f(x) = -x^2 + 1$. If the spreadsheet contains a 1 in cell A1, the formula, -A1^2 +1, returns 2 to any target cell, not 0. The exponentiation is performed on –1 (not 1), as if the equation were $(-x)^2 + 1$. The negation on the left is applied to the value in the cell before the exponentiation. Keep in mind that algebraically, the function could have also been written as $f(x) = 1 - x^2$. The Excel formula, 1 – A1^2, with a 1 in cell A1 returns 0 to any target cell as expected.

Evaluating a Piecewise-Defined Function

4. A government collects a flat 15% income tax on any citizen whose annual taxable income is $50,000 or less. Any citizen whose taxable income exceeds $50,000 but is less than or equal to $200,000 is taxed $7,500 plus 25% of the amount exceeding $50,000. Any citizen whose taxable income exceeds $200,000 is taxed $45,000 plus 30% of the amount exceeding $200,000. Calculate the income tax liability for citizens whose taxable incomes are $24,000, $87,000, and $356,000.

Solution: First write the function that describes the tax law. Let $T(i)$ = the tax amount as a function of the taxable income, i.

$$T(i) = \begin{cases} 0.15i & \text{if } 0 \le i \le 50{,}000 \\ 7{,}500 + 0.25(i - 50{,}000) & \text{if } 50{,}000 < i \le 200{,}000 \\ 45{,}000 + 0.30(i - 200{,}000) & \text{if } i > 200{,}000 \end{cases}$$

Now build a spreadsheet with the three income levels entered into cells A1, A2, and A3. Format the cells as currency. Select cell B1, and enter in the textbox the following formula:

=IF(A1>200000,45000+0.3*(A1-200000), IF(A1>50000,7500+0.25*(A1-50000), 0.15*A1))

The following value is returned for cell B1:

	A	B
1	$24,000.00	$3,600.00
2	$87,000.00	
3	$356,000.00	

Excel always processes the arguments within the "IF" function from left to right. The first argument of the function is evaluated to see if the value of the cell A1 is greater than $200,000. This is false, and so is the next argument. The third argument (an implied "IF") is true because $24,000 is less than or equal to $50,000. Therefore, a 15% tax rate is finally applied. Cell B1 is then evaluated as simply 15% of $24,000 or $3,600. Now drag the lower right-hand corner of cell B1 down to cell B2:

	A	B
1	$24,000.00	$3,600.00
2	$87,000.00	$16,750.00
3	$356,000.00	

For the value in cell A2, the "IF" function is false for the first argument but true for the second argument because the value of A2 is greater than $50,000 but less than or equal to $200,000. Therefore, the 25% tax rate is applied to the amount over $50,000, and cell B2 is evaluated by adding $7,500 + $9,250 = $16,750. Now drag the lower right-hand corner of cell B2 down to cell B3:

	A	B
1	$24,000.00	$3,600.00
2	$87,000.00	$16,750.00
3	$356,000.00	$91,800.00

For the value in cell A3, the "IF" function is true for the first argument, and thus the rest of the statement is ignored. The 30% tax rate is applied to the amount over $200,000, and the total tax is, therefore, $45,000 + $46,800 = $91,800. Note that if the first and third "IF" functions were reversed in the formula, Excel would have ignored the third argument, applied the second argument, and supplied the wrong tax amount. Try it.

1-2 Functions from the Graphical Viewpoint

Graphing Algebraically-Defined Functions

5. Graph the quadratic function $f(x) = x^2 + 2x + 3$ within its domain [0, 30].

Solution: The graph of this function is a curve and will require lots of points to resemble a smooth curve when the points are connected. Suppose we wish to plot 61 points. We could start by typing into column A the values 0, 0.5, 1, ... , 30. That procedure is the brute force way to enter the data. A much easier way is to place a 0 in cell A1 and use a formula for the rest of the values:

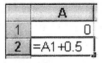

Now click on cell A2 and drag its lower right-hand corner down to cell A61. Excel will quickly supply 61 data values that increase in increments of 0.5, all of which can be used to plot the points:

	A
1	0
2	0.5
3	1
4	1.5

Now click on cell B1 and enter in the formula that describes the function we are trying to plot:

	A	B
1	0	=A1^2+2*A1+3

Then click on cell B1 and drag the lower right-hand corner down to cell B61. Excel will evaluate the function based on all the values in column A and as you drag:

	A	B
1	0	3
2	0.5	4.25
3	1	6
4	1.5	8.25
60	29.5	932.25
61	30	963

Now plot the 61 points using the Chart Wizard. Select all of your data by clicking on column A's top button, hold the Shift key down, and click on column B's top button. Then click on the Chart Wizard icon, ![icon], or select Insert followed by Chart from the main menu. Select the XY (scatter) and "Scatter with data points connected by smoothed lines without markers." chart sub-type. Label your axes x for the horizontal axis and $f(x)$ for the vertical axis. Excel will build the following chart (series box deleted):

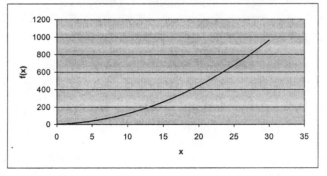

Graphing Piecewise-Defined Functions

6. The average price of a gallon of gasoline in 1999 and 2000 could be estimated with a piecewise linear function of time in months, beginning with the first month of 1999 ($t = 1$) and ending with the sixth month of 2000 ($t = 18$). Suppose the price declined from $1.25 per gallon beginning January 1, 1999 at an average rate of $.02 per month until the end of October, 1999. After October, the price jumped to $1.10 and increased at an average rate of $0.11 per month thereafter. Sketch the graph of the price level of gasoline at the end of each month.

Solution: First write the function that describes the gasoline price trend. Let $P(t)$ be the end-of-month price of gasoline as a function of time in months, t.

$$P(t) = \begin{cases} 1.25 - 0.02t & \text{if } 1 \le t < 11 \\ 1.10 + 0.11(t - 10) & \text{if } t \ge 11 \end{cases}$$

Now build a spreadsheet with the 18 months entered into cells A1, ... , A18. Format the cells in column B as currency. Select cell B1, and enter the formula, =IF(A1<11,1.25-0.02*A1, 1.10+0.11*(A1-10)), into the textbox. The following value is returned for cell B1:

	A	B
1	1	$1.23

Now click on the lower right-hand corner of cell B1 and drag it down to cell 18. Excel will evaluate the function for each of the values in column A:

	A	B
1	1	$1.23
2	2	$1.21
3	3	$1.19
4	4	$1.17
17	17	$1.87
18	18	$1.98

Now plot the 18 points using the Chart Wizard. Select all your data by clicking on column A's top button, hold the Ctrl key down and click on column B's top button. Select the XY (scatter) and "Scatter with data points connected by lines" chart sub-type. Label your axes "Months, t" for the horizontal axis and "Price of Gasoline, $P(t)$" for the vertical axis. Excel will build the following chart (series box deleted):

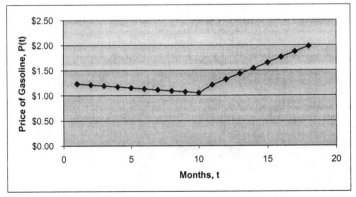

1-3 Linear Functions

Recognizing Linear Data Numerically

7. Consider the following table of values for two variables, x and y.

x	y	x	y	x	y
−7	0	−1	18	5	36
−5	6	0	21	7	42
−4	9	1	24	9	48
−3	12	2	27	10	51
−2	15	3	30	11	54

a) Is this a linear function? Explain why or why not.
b) Write the equation for the line in slope-intercept form.

Solution: To answer the question, we need to determine whether or not the slope of the line is constant. First build a spreadsheet that contains the values of x and y in columns A and B:

	A	B
1	-7	0
2	-5	6
3	-4	9
4	-3	12
14	10	51
15	11	54

Now select cell C1 and enter in the formula for the slope of the line containing the points:

	A	B	C
1	-7	0	=(B2-B1)/(A2-A1)
2	-5	6	

Excel will return a value of 3 for the first two points. Click on the lower right-hand corner of cell C1 and drag it down to cell C14. Excel will calculate the slope of the line between all the successive pairs of values:

	A	B	C
1	-7	0	3
2	-5	6	3
3	-4	9	3
4	-3	12	3
13	9	48	3
14	10	51	3
15	11	54	

The function is indeed linear. A constant slope of 3 exists between every successive pair of values of x and y. Note that a valid slope cannot be obtained for cell C15 because a 16th data point would be required. The y-intercept value is 21, which is the value of y when $x = 0$. Therefore, the equation for the line is $y = 3x + 21$.

The Graph of a Linear Function: Slope and Intercepts

8. Consider the following table of values for two variables, x and y.

x	y	x	y	x	y
−12	−2	2	26	16	54
−8	6	5	32	20	62
−6	10	7	36	24	70
−5	12	9	40	25	72
−1	20	12	46	27	76

Find the x-intercept and the y-intercept by graphing the equation of the line and use these two points to determine the slope.

Solution: Build the spreadsheet with the data values for x in column A and the corresponding data values for y in column B:

	A	B
1	-12	-2
2	-8	6
3	-6	10
4	-5	12
14	25	72
15	27	76

Now use the Chart Wizard to graph the points. Select XY (scatter) and "Scatter with data points connected by lines without markers" for the chart sub-type. Label the horizontal axis x and the vertical axis y. Excel will build the following chart (series box deleted):

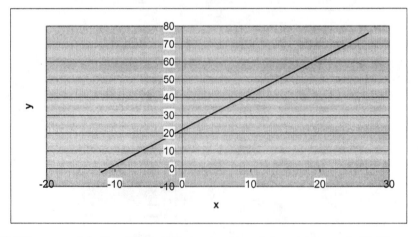

The points all fall on a straight line. The x-intercept is −11 and the y-intercept is 22. If we use these two points to calculate the slope, the result is 22/11 = 2. Note that this value can be checked numerically using successive pairs of points as discussed in problem 7.

1-4 Linear Models

Cost, Revenue, and Profit

9. Carl Blockwood has perfected an exceptionally strong and beautiful way to join wood using a box finger joint manufacturing jig that he designed and built for his table saw. Carl has decided to try to use his joinery skills to build custom toolboxes and small chests from cherry, oak, and walnut hardwoods. Carl estimates his monthly overhead to be $800 (rent, utilities, insurance, etc.) and that his average variable cost to build a box will be $100 per unit. Carl intends to price and sell each box at $200, thus employing the traditional "golden markup"—double the unit cost.
a) What is Carl's break-even quantity each month?
b) What is Carl's total cost of operations if he breaks even?
c) How much profit will Carl earn in a month if he manages to sell 50 boxes?

Solution: Let C = total cost, Q = quantity sold, R = revenue, and P = profit. First write the formula for Carl's total cost and revenue:

$$C = 800 + 100Q$$
$$R = 200Q$$

Profit is simply the difference between revenue and cost:

$$P = R - C = 200Q - (800 + 100Q) = 100Q - 800$$

Now build a spreadsheet that contains various values of Q that realistically cover Carl's scale of operations. Six values from 0 to 50 are reasonable. Enter these into column A:

	A
1	0
2	10
3	20
4	30
5	40
6	50

Now use cell B1 for calculating Carl's total cost at zero volume:

	A	B
1	0	=800+100*A1

Then click on cell B1 and drag the lower right-hand corner down to cell B6 to apply the formula to each quantity value in column A. Format the cells as currency. Next use Column C for calculating Carl's revenue in cell C1:

	A	B	C
1	0	$800.00	=200*A1
2	10	$1,800.00	
3	20	$2,800.00	
4	30	$3,800.00	
5	40	$4,800.00	
6	50	$5,800.00	

Apply the same procedure for cells C1 through C6 as you did for cells B1 through B6. Excel will then calculate the profit for each quantity value in column A. Then use column D for calculating Carl's profit in cell D1:

	A	B	C	D
1	0	$800.00	$0.00	=100*A1 - 800
2	10	$1,800.00	$2,000.00	
3	20	$2,800.00	$4,000.00	
4	30	$3,800.00	$6,000.00	
5	40	$4,800.00	$8,000.00	
6	50	$5,800.00	$10,000.00	

Alternatively, the profit formula in cell D1 could have been C1-B1. The final spreadsheet for all the values in column A is as follows:

	A	B	C	D
1	0	$800.00	$0.00	-$800.00
2	10	$1,800.00	$2,000.00	$200.00
3	20	$2,800.00	$4,000.00	$1,200.00
4	30	$3,800.00	$6,000.00	$2,200.00
5	40	$4,800.00	$8,000.00	$3,200.00
6	50	$5,800.00	$10,000.00	$4,200.00

The calculations reveal that Carl will break even if he sells somewhere between 0 and 10 units each month. We can use the Chart Wizard to zero-in on the break-even point by plotting Carl's total cost, revenue, and profit as a function of quantity. Select all four columns by clicking and dragging the mouse across the top buttons of each column. Select XY (scatter) and "Scatter with data points connected by lines without markers" for the chart sub-type. Label the horizontal axis as Quantity, Series 1 as Total Cost, Series 2 as Revenue, and Series 3 as Profit. Excel will build the following chart:

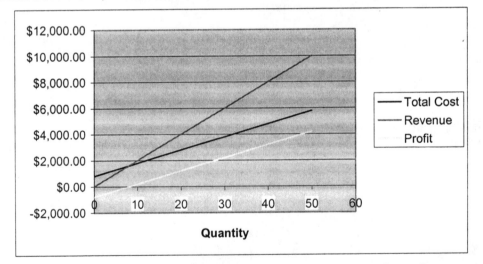

The chart reveals that Carl breaks even (no profit or loss) if he can build and sell 8 toolboxes per month. At break even, his total cost and revenue both equal $1,600. If he could expand his operations enough to build and sell 50 units per month, his monthly profit would be $4,200.

10

1-5 Linear Regression

Computing a Linear Regression Line

10. The following data lists the light output and watts of electricity consumed by 16 different compact fluorescent light bulbs:

Watts	Light Output	Watts	Light Output	Watts	Light Output	Watts	Light Output
15	55	15	65	15	60	27	65
20	60	20	65	16	65	20	80
23	70	30	105	20	80	22	90
28	100	17	60	25	100	30	110

Calculate the slope and intercept of the regression line, assuming that light output is a function of wattage consumed.

Solution: Let x = the watts of electricity consumed by each bulb and y = the light output. The equation for the regression line in slope-intercept form is $y = mx + b$. We need to calculate the slope using the formula, $m = \dfrac{n(\Sigma xy) - (\Sigma x)(\Sigma y)}{n(\Sigma x^2) - (\Sigma x)^2}$, and the y-intercept using the formula, $b = \dfrac{(\Sigma y) - m(\Sigma x)}{n}$, where $n =$ 16. First build a spreadsheet that contains the values for watts in column A and the corresponding light output values in column B:

	A	B
1	15	55
2	20	60
3	23	70
4	28	100
15	22	90
16	30	110

Use column C to find xy values. Select cell C1 and enter in the following formula in the textbox:

	A	B	C
1	15	55	=A1*B1

Excel will multiply the value in A1 by the value in B1. Now select C1 and drag the lower right-hand corner down to cell C16 to perform the same calculations for the rest of the values in columns A and B. Now use column D to find values for x^2. Select cell D1 and enter in the following formula in the textbox:

	A	B	C	D
1	15	55	825	=A1^2

Excel will square the values in A1. Select D1 and drag the lower right-hand corner down to cell D16 to perform the same calculations for the rest of the values in column A. Next select cell A17 and click on

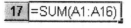

, the AutoSum button, and use it to calculate Σx^2:

17	=SUM(A1:A16)

Now select cell A17 and drag its lower right-hand corner to cell D17. Excel will total all four columns:

17	343	1230	27555	7771

Use the values in row 17 to obtain the slope of the regression line by applying the formula for the slope, m, to cell C18: =(16*C17-A17*B17)/(16*D17-A17^2). That result allows us to apply the formula for the y-intercept to display in cell C19: =(B17-C18*A17)/16. Therefore, the solution to the problem is as follows:

18	Slope of regression line	2.839838
19	y-intercept of regression line	15.99596

Using a Linear Regression Line for Predictions

11. Use the regression model you constructed for problem 10 to predict the light output produced by a:
    a) 25-watt fluorescent bulb
    b) 40-watt fluorescent bulb

Solution: The 25-watt prediction is an interpolation because it falls within the range of the data. On the other hand, the 40-watt prediction is an extrapolation because none of the bulbs shown in the data required this much electricity.

Use cell C21 to perform the interpolation by entering the formula, =25*C18+C19. Then use cell C22 to perform the extrapolation by entering the formula, =40*C18+C19. The solution to each part of the exercise, therefore, is as follows:

21	25-watt interpolation	86.99192
22	40-watt extrapolation	129.5895

The Linear Estimate Function

12. Use the LINEST( ) worksheet function to obtain the slope and intercept for the regression line using the light bulb data presented in problem 10.

Solution: Select any two adjacent cells on the spreadsheet. The left cell will receive the slope, and the right cell will receive the y-intercept. The LINEST( ) function requires that the data for y be specified first and the data for x be specified second. Type the array function, =LINEST(B1:B16,A1:A16), into the textbox. To apply this array function, press Ctrl-Shift-Enter (a keypress combination) so that it will be applied to both cells at the same time. The result is as follows:

24	Slope and intercept found using	Slope	y-intercept
25	the LINEST( ) function:	2.839838	15.99596

Correlation Coefficient

13. Calculate the correlation coefficient for the regression line stemming from the light bulb data presented in problem 10.

Solution: The formula for the correlation coefficient is as follows:

$$r = \frac{n(\Sigma xy) - (\Sigma x)(\Sigma y)}{\sqrt{n(\Sigma x^2) - (\Sigma x)^2} \cdot \sqrt{n(\Sigma y^2) - (\Sigma y)^2}}$$

The only additional calculation we need in order to apply this formula is Σy^2. Use Column E and enter the following formula in the textbox:

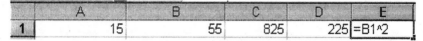

	A	B	C	D	E
1	15	55	825	225	=B1^2

Now select cell E1 and drag its lower right-hand corner down to cell E16 to square all the values of y in column B. Then select cell D17 and click on Σ, the AutoSum icon button to find the total, 99,650. The array formula, =(16*C17-A17*B17)/SQRT(16*D17-A17^2)*SQRT(16*E17-B17^2)), can now be entered into the textbox and applied to another cell in the spreadsheet. Excel will perform the required calculations:

34	Correlation Coefficient =	0.813449

14. Use the LINEST( ) function to calculate the correlation coefficient for the light bulb data presented in problem 10.

Solution: The LINEST( ) worksheet function will return the slope, y-intercept, coefficient of determination, and several other statistics. The correlation coefficient is the square root of the coefficient of determination. The sign of the correlation coefficient is the same as the slope. Select a block of empty cells two wide and five high. Then type the array function, =LINEST(B1:B16,A1:A16,TRUE,TRUE), in the textbox. Press Ctrl-Shift-Enter to apply the array function to the data. Excel returns the values to the block:

29	Statistics found using the LINEST( ) function	2.839838	15.99596
30			0.542688	11.95993
31			0.6617	11.09445
32			27.38336	14
33			3370.533	1723.217

The coefficient of determination, 0.6617, is shown in cell D31, and the slope is positive. Therefore, apply the formula, =SQRT(D31), to cell C34, and Excel returns:

34	Correlation Coefficient =	0.813449

Graphing a Regression Line

15. Graph the regression line together with a scatter plot of the data presented in problem 10.

Solution: Select the data in cells A1 to A16 and B1 to B6. Use the Chart Wizard by selecting XY (scatter) and "Compares pairs of values" for the chart sub-type. Choose the Series in columns option, and label your axes "Wattage" for variable x and "Light Output" for variable y. After Excel displays the chart, delete the Series box and select Chart from the main menu. Then select the Add Trendline, and choose the Linear Trend/Regression type. Excel will add the trendline to the scatter plot to produce the chart on the next page:

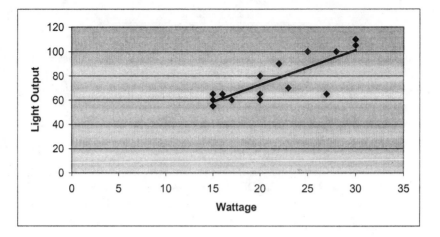

Chapter Review Questions

1. Let f be the quadratic function defined by $f(x) = 3x^2 + 5x - 17$ with domain $(0, 300]$. Evaluate the function for values $x = 6, \ldots, 15$.

2. The ABC real estate brokerage firm pays its salespeople using a compensation plan consisting of three plateaus based on the level of property sales. All sales employees qualify for Plateau A: a 5% commission is paid to all salespeople for the first $1 million in sales volume that each accumulates during a given year. The more successful salespeople qualify for Plateau B: if their volume exceeds $1 million but is less than or equal to $5 million, ABC pays $50,000 plus 6% of the amount exceeding $1 million. The top salespeople qualify for Plateau C: any salesperson whose sales volume exceeds $5 million is paid $290,000 plus 7% of the amount exceeding $5 million. Write the piecewise-defined function and calculate the annual commissions paid to three salespeople whose annual sales volumes are as follows: Alice, $750,000; Bob, $3.6 million; and Charles, $9.2 million.

3. Sketch the graph of the piecewise-defined function for the compensation plan offered by the ABC real estate brokerage firm. Build your graph by selecting five levels of sales volume that fall within each of the three plateaus.

4. Consider the following table of values for two variables, x and y.

x	y	x	y
−2	0	8	25
0	5	10	30
2	10	12	35
4	15	14	40
6	20	16	45

a) Is this a linear function? Explain why or why not.
b) Write the equation for the line in slope-intercept form.

5. Alex Hamilton has decided to employ his woodturning craftsmanship to build drumsticks using imported exotic hardwoods—bubinga, cocobola, gaboon, and rosewood. Alex is an excellent drummer as well and will use the back room of his music studio as a workshop to start the new business, thus keeping overhead costs at a bare minimum. Alex estimates his monthly overhead for the new workshop to be $300 (rent, utilities, insurance, etc.) and that his average variable cost to build a pair of drumsticks will be $20 per pair. Alex intends to price and sell each pair of drumsticks at an average of $50.
   a) What is Alex's break-even quantity each month?
   b) What is Alex's total cost of operations if he breaks even?
   c) How much profit will Alex earn in a month if he manages to sell 30 pairs of drumsticks?
   d) Sketch the graph of Alex's cost, revenue, and profit functions.

6. The following data reveal the weights (lb) of 12 sport utility vehicles together with their estimated overall fuel economy (mpg):

Weight	Fuel Economy	Weight	Fuel Economy	Weight	Fuel Economy
2500	26.1	4300	13.4	2600	25.6
2550	24.2	3100	19.5	2900	23.7
2700	22.4	4800	11.0	3600	19.1
3500	18.3	3400	20.1	4000	17.2

   a) Calculate the slope and intercept of the regression line, assuming that fuel economy is a function of vehicle weight. Use Excel's LINEST( ) function.
   b) Use your regression equation to predict the fuel economy of an SUV that weighs 3900 lb.
   c) Calculate the correlation coefficient for the two variables.
   d) Graph the regression line together with a scatter plot.

CHAPTER 2

Non-Linear Functions and Models

2-1 Quadratic Functions and Models

Graphing Quadratic Functions

1. Sketch the graph of $f(x) = 3.4x^2 - 7.8x - 5.5$ and determine if it is concave upward or downward, locate the vertex, y-intercept, and the x-intercepts. Also find the line of symmetry.

Solution: The numbers -5, -4.5, ... , 7.5 are entered into the Excel spreadsheet from A1 to A26. The function is evaluated in B1 as shown in the following figures.

	A	B	C
1	-5	=3.4*A1^2-7.8*A1-5.5	
2	-4.5	98.45	
3	-4	80.1	
4	-3.5	63.45	
5	-3	48.5	

$\Longrightarrow$

	A	B
1	-5	118.5
2	-4.5	98.45
3	-4	80.1
4	-3.5	63.45
5	-3	48.5

After executing a click–and–drag in column B from B1 to B26, the function values are obtained for the values of x in column A. A part of the results are shown above.

The Chart Wizard is now chosen and the chart type XY (scatter) is selected to obtain the following plot.

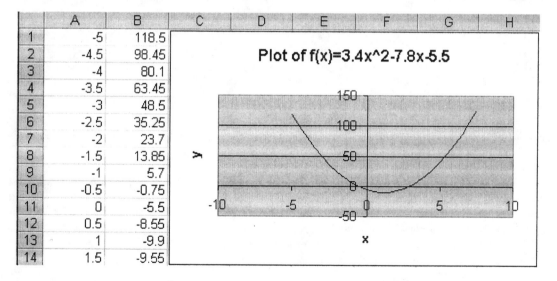

	A	B
1	-5	118.5
2	-4.5	98.45
3	-4	80.1
4	-3.5	63.45
5	-3	48.5
6	-2.5	35.25
7	-2	23.7
8	-1.5	13.85
9	-1	5.7
10	-0.5	-0.75
11	0	-5.5
12	0.5	-8.55
13	1	-9.9
14	1.5	-9.55

Plot of f(x)=3.4x^2-7.8x-5.5

The Excel plot for $f(x) = 3.4x^2 - 7.8x - 5.5$, shows that the curve is concave upward. The vertex is located at $(x, y) = (1.147, -9.973)$, the y-intercept is -5.5, and the x-intercepts are -0.566 and 2.860. The line of symmetry is $x = 1.147$.

2. Sketch the graph of $f(x) = -1.075x^2 + 12.005x - 2.375$ and determine if it is concave upward or downward, locate the vertex, y-intercept, and the x-intercepts. Also find the line of symmetry.

Solution: Enter a set of convenient numbers, −3 through 15, into column A for values of x. Then evaluate the function in column B as follows:

	A	B	C	D
1	-3	=-1.075*A1^2+12.005*A1-2.375		
2	-2	-30.685		
3	-1	-15.455		
4	0	-2.375		
5	1	8.555		

The Chart Wizard is now chosen and the chart type XY (scatter) is selected to obtain the following plot.

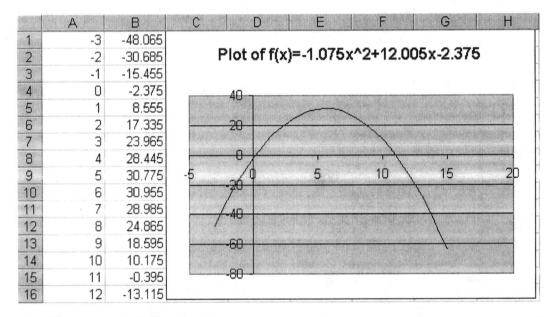

	A	B
1	-3	-48.065
2	-2	-30.685
3	-1	-15.455
4	0	-2.375
5	1	8.555
6	2	17.335
7	3	23.965
8	4	28.445
9	5	30.775
10	6	30.955
11	7	28.985
12	8	24.865
13	9	18.595
14	10	10.175
15	11	-0.395
16	12	-13.115

Plot of f(x)=-1.075x^2+12.005x-2.375

This parabola is concave downward with the vertex located at $(x, y) = (5.584, 31.141)$. The y-intercept is −2.375, and the x-intercepts are 0.201 and 10.966. The line of symmetry is $x = 5.584$.

3. MidwestUnivResearch.edu is an Internet website that describes ongoing research projects at Midwest University. One such research project concerns the relationship between fertilizer application and wheat yield. Several different levels of fertilizer, x, are applied to the same size plots and the yield of wheat in bushels per plot, y, is measured. The function relating yield to fertilizer level is $f(x) = -x^2 + 10x + 30$. Plot the wheat yield versus fertilizer level, determine the optimum amount of fertilizer, and the wheat yield when this optimum level is applied.

Solution: Using a similar procedure described in problem 2, a set of convenient numbers is first entered into column A for values of x. The formula, -A^2+10*A+30, is applied to cells B1:B11 to evaluate the function. The points are then plotted using the Chart Wizard:

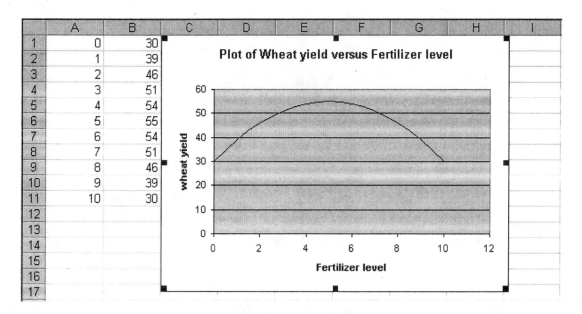

	A	B
1	0	30
2	1	39
3	2	46
4	3	51
5	4	54
6	5	55
7	6	54
8	7	51
9	8	46
10	9	39
11	10	30

Plot of Wheat yield versus Fertilizer level

The maximum yield occurs at the vertex, (5, 55). Note that wheat yield increases as the fertilizer level is increased until the level $x = 5$ is reached. Increasing the level beyond 5 units per plot decreases yield.

2-2 Exponential Functions and Models

Exponential Functions from the Numerical Point of View

4. Comparing simple interest with compound interest provides an interesting contrast between linear functions and exponential functions. In simple interest, the amount of interest earned each year is constant. In other words, only the original principal earns interest from year to year, and the interest earned in any given year does not earn interest in future years. In compound interest, the interest earned in one year earns interest itself in future years. Consider two different scenarios where $5,000 is invested at 7% simple interest and $5,000 is invested at 7% compound interest. Compare the amounts present after t years for $t = 1$ through 10 for both types of interest and compute the difference between the two.

Solution: The accumulation functions for these two cases and the Excel spreadsheet are as follows:

Simple Interest: $S(t) = 5,000 + 5,000(0.07)t = 5,000 + 350t$

Compound Interest: $C(t) = 5,000(1.07)^t$

Use row 1 for table headings and column A for values of t from 1 through 10. Then enter the simple interest formula, =5000 + 350*A2, into cell B2 and the compound interest formula, =5000*1.07^A2, into cell C2. The difference is found in cell D2 using the formula, =C2-B2. Following the execution of a click-and-drag in columns B, C, and D, we obtain the following output.

	A	B	C	D
1	years	simple interest	compound interest	difference
2	1	5350	5350	0
3	2	5700	5724.5	24.5
4	3	6050	6125.215	75.215
5	4	6400	6553.98005	153.9801
6	5	6750	7012.758654	262.7587
7	6	7100	7503.651759	403.6518
8	7	7450	8028.907382	578.9074
9	8	7800	8590.930899	790.9309
10	9	8150	9192.296062	1042.296
11	10	8500	9835.756786	1335.757

The simple interest function is a linear function. The slope is 350. For an increase of 1 year, the amount increases by 350. The compound interest function is an exponential function. For an increase of 1 year, multiply the amount by 1.07.

Exponential Functions from the Graphical Point of View

5. Graphically compare simple interest (a linear function) with compound interest (an exponential function).

Solution: The data from problem 4 is plotted using the Chart Wizard. Series 1 shows simple interest and series 2 shows compound interest.

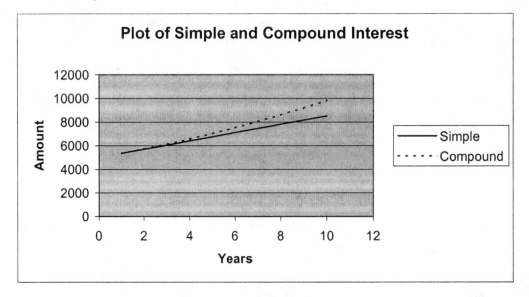

From the graphs, the difference between linear and exponential growth is clear. The spread between the two functions increases as time goes by.

20

Finding an Exponential Equation from Data: How to Make an Exponential Model

6. The cost of a funeral seems to be rising exponentially. The following table gives the average cost of a funeral for some years in the recent past:

Year	Average Cost
1980	$1,809
1985	$2,737
1991	$3,742
1995	$4,624
1996	$4,782
1998	$5,020

Predict the average cost in 2002.

Solution: We code the years in the table as 1, 6, 12, 16, 17, and 19. The coded years are entered in column A, and the average costs are entered in column B. Using the Chart Wizard and the chart type XY(scatter), the six points are plotted. The chart menu is selected and Add Trendline is selected. In the Add Trendline dialog box, select exponential type, and under options, select Display equation on chart.

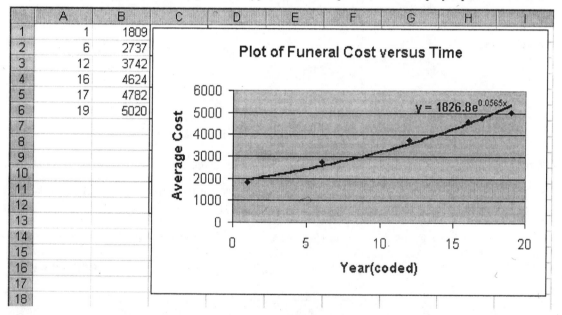

The coded value for the year 2002 is 23. Evaluating the exponential equation for $x = 23$, we find the following estimate for the average cost of a funeral in 2002. This estimate assumes that the exponential model will describe the cost in 2002.

$$1826.8e^{0.0565(23)} = \$6,699.7$$

The Number e and Further Applications

7. Investigate the behavior of the following expression for increasing values of m.

$$\left(1+\frac{1}{m}\right)^m.$$

Solution: The powers of ten from 1 through 100,000,000 are entered from A1 through A9. Enter a 1 into cell A1, and use the formula, =10*A1, in cell A2. Then click and drag cell A2 to cell A9. Now enter the formula, =(1+1/A1)^A1, into cell B1, and click and drag cell B1 to cell B9 to return the required values.

This exercise illustrates that as m increases, the expression $\left(1+\frac{1}{m}\right)^m$ approaches a constant value. In calculus, this constant value is represented by the letter e, and is equal to 2.718282 to six decimal places.

	A	B
1	1	2
2	10	2.593742
3	100	2.704814
4	1000	2.716924
5	10000	2.718146
6	100000	2.718268
7	1000000	2.71828
8	10000000	2.718282
9	100000000	2.718282

8. If an amount (present value) P is invested for t years at an annual rate of r, and if the interest is compounded m times per year, then the future value A is given by $A(t)=P\left(1+\frac{r}{m}\right)^{mt}$. Find the future value of $1,000 invested at an annual rate equal to 5% if the interest is compounded yearly, monthly, weekly, daily, hourly, each minute, and each second in a non leap year.

Solution: One year is composed of 12 months, or 52 weeks, or 365 days, or 8,760 hours, or 525,600 minutes, or 31,536,000 seconds in a year. Enter these values into column A. Then enter the formula, =1000*(1+0.05/A1)^A1, into cell B1. After a click-and-drag is performed in column B, Excel returns the following values:

	A	B
1	1	1050
2	12	1051.161898
3	52	1051.245842
4	365	1051.267496
5	8760	1051.270946
6	525600	1051.271094
7	31536000	1051.271093

If P dollars are invested at an annual interest rate r compounded continuously, the accumulated amount after t years is $A(t) = Pe^{rt}$. In this example, $A(1) = 1000e^{0.05(1)} = 1000(1.051271096) = 1051.271$. As shown in the Excel output, as the frequency of compounding increases, the amount present approaches $1000e^{0.05(1)}$. Notice that the difference between minutes and seconds in the amount of compounded interest on \$1,000 does not appear until six digits to the right of the decimal place.

2-3 Logarithmic Functions and Models

Evaluating Logarithms Using Excel

9. In Excel, log(number, base) calculates the logarithm of any number you supply to the base you supply, ln(number) calculates the natural logarithm of any number you supply, and log10(number) calculates the common or base ten logarithm of any number you supply.

Solution: The following Excel output illustrates the use of these functions.

	A	B	C	D	E
1	numbers	natural log	common log	powers of two	log to base 2 of number in column D
2	1	0	0	2	1
3	2	0.693147	0.301029996	4	2
4	3	1.098612	0.477121255	8	3
5	4	1.386294	0.602059991	16	4
6	5	1.609438	0.698970004	32	5
7	6	1.791759	0.77815125	64	6
8	7	1.94591	0.84509804	128	7
9	8	2.079442	0.903089987	256	8
10	9	2.197225	0.954242509	512	9
11	10	2.302585	1	1024	10

Enter the numbers 1 through 10 into cells A2:A11. Then enter the formula, =LN(A2), into cell B2 in order to obtain natural logs. Then enter the formula, =LOG10(A2), into cell C2 in order to obtain logs to the base 10. A click-and-drag then returns the results in columns B and C. Enter a 2 into Cell D2 and use the formula, 2*D2, for cell D3. Click and drag cell D3 to cell D11 to return the first ten powers of two. Finally the formula, =LOG(D2,2), is used to obtain logs to the base 2 and is entered into cell E2. A click-and-drag produces the output shown in column E. Note that the common logs could also have been obtained using the formula, LOG(A2,10).

Graphs of Logarithm Functions

10. Compare the natural and common logarithm functions for domain values between 0 and 1 inclusive.

Solution: The following Excel output compares the two functions using five convenient values for x where $0 < x \le 1$ in column A. The natural log of each value is calculated in column B and plotted as a solid line curve. The common log of each value is calculated in column C and plotted as a dashed line curve:

	A	B	C
1	x	ln(x)	log10(x)
2	0.01	-4.60517	-2
3	0.05	-2.99573	-1.30103
4	0.1	-2.30259	-1
5	0.5	-0.69315	-0.30103
6	1	0	0
7			
8			
9			
10			
11			
12			
13			
14			

Plot of Common and Natural Logarithm Functions

11. Compare the natural and common logarithm functions for domain values greater than or equal to 1.

Solution: The following Excel output compares the two functions using five values of x that range from 1 to 20. The natural log of each value is calculated in column B and plotted as a solid line curve. The common log of each value is calculated in column C and plotted as a dashed line curve:

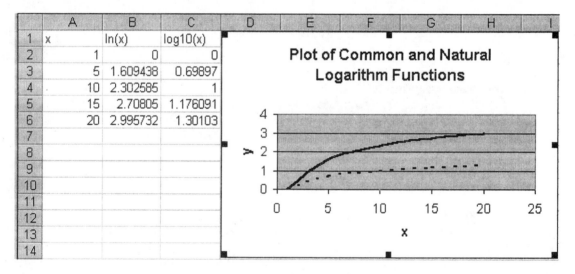

	A	B	C
1	x	ln(x)	log10(x)
2	1	0	0
3	5	1.609438	0.69897
4	10	2.302585	1
5	15	2.70805	1.176091
6	20	2.995732	1.30103
7			
8			
9			
10			
11			
12			
13			
14			

Plot of Common and Natural Logarithm Functions

Algebraic Properties of Logarithms

12. Find and compare the following natural logarithms: Compare $\ln(2 \times 3)$ with $\ln(2) + \ln(3)$, Compare $\ln\left(\dfrac{3}{2}\right)$ with $\ln(3) - \ln(2)$, and Compare $\ln(2^3)$ with $3 \ln(2)$.

Solution: Use Excel's LN( ) function to return natural logs. Select an empty cell and enter the formula, =LN(2*3). Excel returns 1.791759. In similar fashion, the formula, =LN(2)+LN(3), returns 1.791759; =LN(3/2) returns 0.405465; =LN(3)-LN(2) returns 0.405465; =LN(2^3) returns 2.079442; and =3*LN(2) returns 2.079442.

2-4 Trigonometric Functions and Models

The Sine Function

13. Plot the sine function, $y = \sin(x)$, for values of x ranging from 0 to 15 radians.

Solution: Excel's SIN( ) function returns the sine of an angle measured in radians. Enter the values of x from 0 to 15, at intervals of 0.5, into column A. Then enter the formula, =SIN(A1), into cell B1. Execute a click-and-drag in column B to obtain values of y. Finally, use the Chart Wizard to produce the following graph of the sine function for x ranging from 0 to 15 radians:

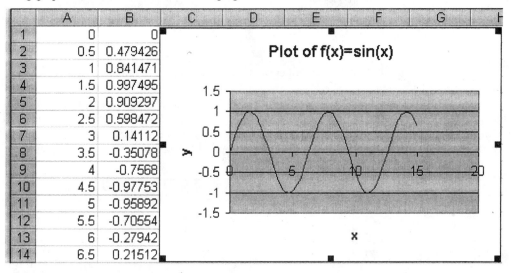

Note that the sine curve has amplitude equal to 1 and period equal to $2\pi \approx$ approximately 6.28.

14. Plot the function $y = 3 \sin[2(x - 3.14)] + 1$.

Solution: The general sine function is $f(x) = A\sin[\omega(x - \alpha)] + C$, where A is the amplitude, C is the vertical offset, ω is the angular frequency, α is the phase shift, and $P = 2\pi/\omega$ is the period or wavelength. For the function $y = 3 \sin[2(x - 3.14)] + 1$, $A = 3$, $\omega = 2$, $\alpha = 3.14$, and $C = 1$. Obtain incremental values for x by entering 0 into cell A1 and applying the formula, =A1+.25, to cell A2. Fill more values in column A by using a click-and-drag operation. Then enter the formula, =3*SIN(2*(A1-3.14))+1, into cell B1. A click-and-drag returns the required points on the curve, ready for plotting. Finally, use the Chart Wizard to plot the graph of the function. The graph oscillates between $C - A = -2$ and $C + A = 4$. The phase shift is $\alpha = 3.14$. The period is $P = 2\pi/2 = \pi$. Since $\omega = 2$, the function completes two cycles for x between 0 and 2π.

	A	B	C	D	E	F	G	H
1	0	1.009555905						
2	0.25	2.446655415			Plot of f(x)=3sin(2(x-3.14))+1			
3	0.5	3.529563226						
4	0.75	3.993145737						
5	1	3.723901782						
6	1.25	2.787751671						
7	1.5	1.413897602						
8	1.75	-0.061293036						
9	2	-1.276642125						
10	2.25	-1.93458982						
11	2.5	-1.874047581						
12	2.75	-1.109838258						
13	3	0.170933054						
14	3.25	1.654688869						

The Cosine Function

15. Plot the cosine function $y = \cos(x)$ for values of x ranging from 0 to 15 radians.

Solution: Excel's COS( ) function returns the cosine of an angle measured in radians. Enter the values of x from 0 to 15, at intervals of 0.5, into column A. (Use a procedure similar to that used for the previous problem.) Enter the formula, =COS(A1), into cell B1, and execute a click-and-drag in column B. Finally, use the Chart Wizard to produce the following graph of the sine function for x ranging from 0 to 15 radians:

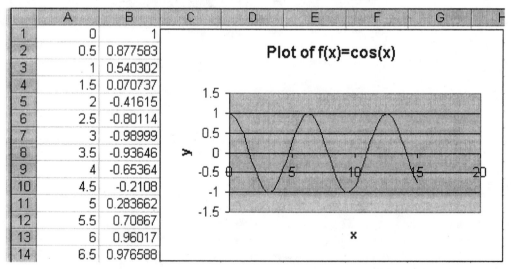

	A	B	C	D	E	F	G	H
1	0	1						
2	0.5	0.877583			Plot of f(x)=cos(x)			
3	1	0.540302						
4	1.5	0.070737						
5	2	-0.41615						
6	2.5	-0.80114						
7	3	-0.98999						
8	3.5	-0.93646						
9	4	-0.65364						
10	4.5	-0.2108						
11	5	0.283662						
12	5.5	0.70867						
13	6	0.96017						
14	6.5	0.976588						

16. Plot the function $y = 3\cos(2(x - 3.14)) + 1$.

Solution: The general cosine function is $f(x) = A\cos[\omega(x - \alpha)] + C$, where A is the amplitude, C is the vertical offset, ω is the angular frequency, α is the phase shift, and $P = 2\pi/\omega$ is the period or wavelength.

For the function, $y = 3 \cos(2(x - 3.14) + 1$, $A = 3$, $\omega = 2$, $\alpha = 3.14$, and $C = 1$. Values of x at intervals of 0.25 are obtained for column A. The formula, =3*COS(2*(A1-3.14))+1, is entered into cell B1 and a click-and-drag calculates the points on the curve. Now use the Chart Wizard to produce the graph of the function:

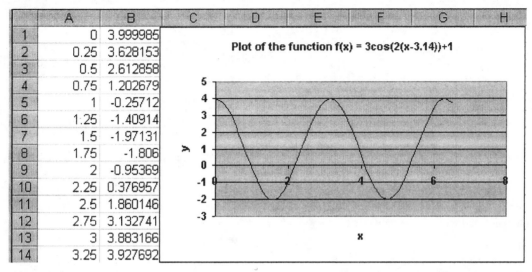

The graph oscillates between $C - A = -2$ and $C + A = 4$. The phase shift is $\alpha = 3.14$. The period is $P = 2\pi/2 = \pi$. Since $\omega = 2$, the function completes two cycles for x between 0 and 2π.

The Tangent Function

17. Investigate the behavior of the function $y = \tan(x)$.

Solution: The plotting of the other four trigonometric functions presents difficulties not experienced with the sine and cosine functions. The tangent, cotangent, secant, and cosecant functions all have asymptotes. The function values approach infinity or minus infinity near these asymptotes. This makes it difficult to accurately graph these functions using computer software because of the scaling needed to accommodate their behavior near the asymptotes. This will be illustrated by the Excel spreadsheet that accompanies this problem for the tangent function values for $0 \le x \le \pi$. For values of x near but less than $\pi/2$, the values of the tangent function are positive and very large. On the other hand, for values of x near but greater than $\pi/2$, the values of the tangent function are negative and are very large in absolute value.

Begin by entering the sample of x values in columns A, C, E, and G into the Excel spreadsheet. Excel's TAN( ) function returns the tangent of an angle measured in radians. Therefore, enter the following formulas: =TAN(A2) in cell B2; =TAN(C2) in cell D2; =TAN(E2) in cell F2; and =TAN(G2) in cell H2. The click-and-drag operations in columns B, D, F, and H return the output shown in the following panel:

	A	B	C	D	E	F	G	H
1	x	tan(x)	x	tan(x)	x	tan(x)	x	tan(x)
2	0	0	1	1.557408	1.57079633	-312002409	2.3	-1.11921
3	0.1	0.100335	1.1	1.96476	1.5708	-272241.81	2.4	-0.91601
4	0.2	0.20271	1.2	2.572152	1.58	-108.6492	2.5	-0.74702
5	0.3	0.309336	1.3	3.602102	1.6	-34.232533	2.6	-0.6016
6	0.4	0.422793	1.4	5.797884	1.7	-7.6966021	2.7	-0.47273
7	0.5	0.546302	1.5	14.10142	1.8	-4.2862617	2.8	-0.35553
8	0.6	0.684137	1.56	92.6205	1.9	-2.9270975	2.9	-0.24641
9	0.7	0.842288	1.57	1255.766	2	-2.1850399	3	-0.14255
10	0.8	1.029639	1.5707	10381.33	2.1	-1.7098465	3.14159265	-3.6E-09
11	0.9	1.260158	1.57079632	1.47E+08	2.2	-1.3738231	3.14159266	6.41E-09

As the values of x get closer to $\pi/2$, the values of the tangent function get larger and larger. The number 1.57079632 is smaller than $\pi/2$ and the value of the tangent function is 1.47E+08 or 147,000,000. The number 1.57079633 is greater than $\pi/2$ and the value of the tangent function is −312,002,409. As the values of x increase from $\pi/2$ to π, the values of the tangent function are negative, but decrease in absolute value and when $x = \pi$, $\tan(x) = 0$.

Chapter Review Exercises

1. Plot the graph of the function, $f(x) = -x^2 + 4x - 4$, for x values between −5 and 5. Discuss concavity, intercepts, the vertex, and the line of symmetry.

2. The proportion of a population that subscribes to a new Internet service after it is advertised x times is given by $P(x) = 0.65(1 - e^{-0.05x})$. Plot this exponential function. As the number of advertisements increases, what limiting percent of the population does the number of subscribers approach?

3. Use the logarithm property, $\log_b x = \dfrac{\log_a x}{\log_a b}$, to evaluate $\log_3 x$ for $x = 1, 2, 3, \ldots, 25$ and use these values to graph $y = \log_3 x$.

4. Plot $y = 0.25 \sin[0.5(x + 3.14)] - 2$ for selected x values between 0 and 15.

5. Build a table of values for $y = \cot(x)$ for $0 < x < \pi$. Be sure and choose some values near 0 and π to clearly illustrate the asymptotic behavior of the cotangent function.

CHAPTER 3

Introduction to the Derivative

3-1 Average Rate of Change

Numerical Point of View

1. Statistical Services is a consulting company that made profits shown in the following table:

Year	Profits ($000s)
1997	75
1998	88
1999	97
2000	108

    a) Find the average rate of change in profits from 1997 to 1998, 1999, and 2000.

    b) Find the rate of change in profits from year to year.

Solution: For part a), build a spreadsheet containing the headings and data covering the years and profits using cells A1:B5. Then enter the formula, =(B3-B$2)/(A3-A$2), into cell C3. A click-and-drag operation returns the values for =(B4-B$2)/(A4-A$2) in cell C4 and =(B5-B$2)/(A5-A$2) in cell C5. Excel returns the average rate of change in profits:

	A	B	C
1	year	profit	average rate of change
2	1997	75	
3	1998	88	13
4	1999	97	11
5	2000	108	11

For part b), enter the formula, =(B3-B2)/(A3-A2), into cell C3. Adjust the cell C1 heading to "rate of change". A click-and-drag results in the value for =(B4-B3)/(A4-A3) returned to cell C4 and the value for =(B5-B4)/(A5-A4) returned to cell C5—the rate of change in profits from year to year:

	A	B	C
1	year	profit	rate of change
2	1997	75	
3	1998	88	13
4	1999	97	9
5	2000	108	11

Graphical Point of View

2. The following Excel graph shows the yearly profits for the years 1995 through 2000 for Books.com:

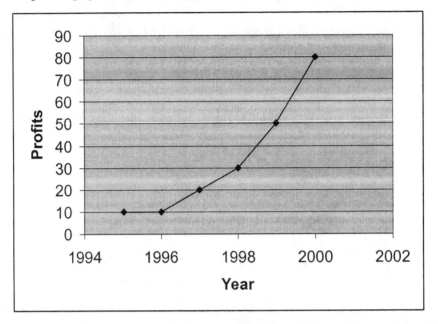

By considering the graph, we see that the average yearly rates of change are as follows: 0 from 1995 to 1996; 10 from 1996 to 1997; 10 from 1997 to 1998; 20 from 1998 to 1999; and 30 from 1999 to 2000. Confirm these rates of change using Excel.

Solution: The average yearly rates of change using Excel are as follows. In cell C3, enter the formula, =(B3-B2)/(A3-A2), and then perform a click-and-drag. Excel returns the results in cells C3 through C7. These are the same results that are found by considering the graph of profits versus years.

	A	B	C
1	Year	Profit	Yearly Rate of Change
2	1995	10	
3	1996	10	0
4	1997	20	10
5	1998	30	10
6	1999	50	20
7	2000	80	30

Algebraic Point of View

3. The growth of CD sales from 1990 through 1999 is described by $S = 254 + 99t - 3t^2$, where $t = 0$ corresponds to 1990 and $t = 9$ corresponds to 1999. The year 1995 corresponds to $t = 5$. Investigate the average rate of change from 5 to $5 + h$, when $h = 1, 0.1, 0.01, 0.001,$ and 0.0001.

Solution: Enter the values for h into column A. Then enter the formula for the difference quotient in cell B1 for h: =((254+99*(5+A1)-3*(5+A1)^2)-(254+99*5-3*5^2))/A1. Next, copy the content of cell B1 into cells B2:B5 using a click-and-drag operation. Excel returns the following results.

	A	B
1	1	66
2	0.1	68.7
3	0.01	68.97
4	0.001	68.997
5	0.0001	68.9997

3-2 The Derivative as Rate of Change: A Numerical Approach

4. Use the numerical approach to find the derivative of the natural log function, $f(x) = \ln(x)$, when $x = 1$.

Solution: The derivative of $\ln(x)$ when $x = 1$ is defined as the limit as h approaches 0 of the quotient $\dfrac{\ln(1 + h) - \ln(1)}{h}$. Enter the positive h values 1, 0.1, 0.01, 0.001, and 0.0001 into column A, and enter the negative h values -0.9, -0.1, -0.01, -0.001, and -0.0001 into column C. Evaluate the function by entering the formula, =(LN(1+A1)-LN(1))/A1, into cell B1. Then copy the contents of cell B1 into cells B2 through B5. Perform a similar operation for cells D1 through D5:

	A	B	C	D
1	1	=(LN(1+A1)-LN(1))/A1	-0.9	=(LN(1+C1)-LN(1))/C1
2	0.1	=(LN(1+A2)-LN(1))/A2	-0.1	=(LN(1+C1)-LN(1))/C2
3	0.01	=(LN(1+A3)-LN(1))/A3	-0.01	=(LN(1+C1)-LN(1))/C3
4	0.001	=(LN(1+A4)-LN(1))/A4	-0.001	=(LN(1+C1)-LN(1))/C4
5	0.0001	=(LN(1+A5)-LN(1))/A5	-0.0001	=(LN(1+C1)-LN(1))/C5

When the formulas are executed, Excel returns the following values for the quotient:

	A	B	C	D
1	1	0.693147181	-0.9	2.558427881
2	0.1	0.953101798	-0.1	1.053605157
3	0.01	0.995033085	-0.01	1.005033585
4	0.001	0.999500333	-0.001	1.000500334
5	0.0001	0.999950003	-0.0001	1.000050003

The values of the quotient $\dfrac{\ln(1 + h) - \ln(1)}{h}$ approach 1 as h approaches 0 from both directions. The derivative of $f(x) = \ln(x)$ at $x = 1$ is equal to 1. Note that in cell C1, we started at $h = -0.9$ rather than -1 because using -1 would require the evaluation of $\ln(0)$, which is undefined.

3-3 The Derivative as Slope: A Geometric Approach

5. Graph the function $f(x) = x^3$ for x between -1 and 1. Evaluate the slope of the secant line connecting the points $(-h, f(-h))$ and $(h, f(h))$ for $h = 1$, 0.1, 0.01, 0.001, and 0.0001. What is the slope of the tangent line to the curve $f(x) = x^3$ at the point $(0, 0)$? What is the connection between the slopes of the secant lines and the slope of the tangent line at $(0, 0)$?

Solution: The graph of $f(x) = x^3$ is shown in the following worksheet by using the Chart Wizard to plot the values in columns A and B:

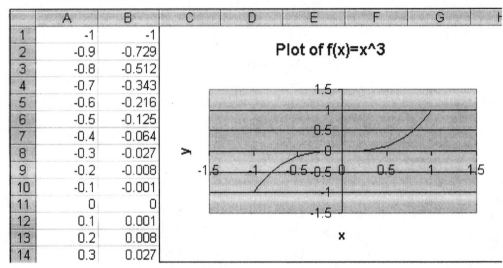

	A	B	C	D	E	F	G	H
1	-1	-1						
2	-0.9	-0.729						
3	-0.8	-0.512						
4	-0.7	-0.343						
5	-0.6	-0.216						
6	-0.5	-0.125						
7	-0.4	-0.064						
8	-0.3	-0.027						
9	-0.2	-0.008						
10	-0.1	-0.001						
11	0	0						
12	0.1	0.001						
13	0.2	0.008						
14	0.3	0.027						

Plot of f(x)=x^3

Note that it is symmetric about the origin. The curve rises from $(-1, -1)$ to $(1, 1)$.

The slope of the secant line connecting the points $(-h, f(-h))$ and $(h, f(h))$ is given by $\dfrac{f(h) - f(-h)}{2h}$.

The following spreadsheet shows the evaluation of this function for $h = 1$, 0.1, 0.01, 0.001, and 0.0001. These values are entered into column A. The required formulas are shown followed by the returned values in column B:

	A	B
1	1	=(A1^3-(-A1)^3)/(2*A1)
2	0.1	=(A2^3-(-A2)^3)/(2*A2)
3	0.01	=(A3^3-(-A3)^3)/(2*A3)
4	0.001	=(A4^3-(-A4)^3)/(2*A4)
5	0.0001	=(A5^3-(-A5)^3)/(2*A5)

	A	B
1	1	1
2	0.1	0.01
3	0.01	0.0001
4	0.001	0.000001
5	0.0001	0.00000001

Note that the slopes of the secant lines are approaching 0 as h approaches 0. In fact, the slope of the tangent line at $h = 0$ is 0. This is also seen by examining the graph of the function, where the slope of the tangent line is zero at the origin.

3-4 The Derivative as a Function: An Algebraic Approach

Functions Not Differentiable at a Point

6. The function $f(x) = \sqrt[5]{x} = x^{1/5}$ has derivative $\dfrac{1}{5}x^{-4/5}$. This derivative is undefined at $x = 0$. Plot

thegraph of the function and investigate the behavior of the $\dfrac{f(h) - f(-h)}{2h}$ as h approaches 0.

Solution: The required spreadsheet and graph are as follows:

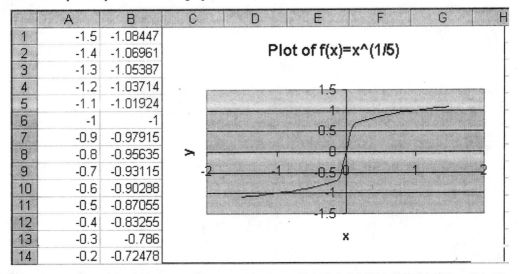

	A	B	C	D	E	F	G	H
1	-1.5	-1.08447						
2	-1.4	-1.06961						
3	-1.3	-1.05387						
4	-1.2	-1.03714						
5	-1.1	-1.01924						
6	-1	-1						
7	-0.9	-0.97915						
8	-0.8	-0.95635						
9	-0.7	-0.93115						
10	-0.6	-0.90288						
11	-0.5	-0.87055						
12	-0.4	-0.83255						
13	-0.3	-0.786						
14	-0.2	-0.72478						

Plot of f(x)=x^{1/5}

Now enter the h values into column A and the expression =(A1^0.2-(-A1)^0.2)/(2*A1) into B1. Then copy the contents of cell B1 into cells B2 through B5. The following is obtained:

	A	B
1	1	=(A1^0.2-(-A1)^0.2)/(2*A1)
2	0.1	=(A2^0.2-(-A2)^0.2)/(2*A2)
3	0.01	=(A3^0.2-(-A3)^0.2)/(2*A3)
4	0.001	=(A4^0.2-(-A4)^0.2)/(2*A4)
5	0.0001	=(A5^0.2-(-A5)^0.2)/(2*A5)

$\Longrightarrow$

	A	B
1	1	1
2	0.1	6.30957344
3	0.01	39.8107171
4	0.001	251.188643
5	0.0001	1584.89319

As h approaches 0, the value of the derivative becomes larger and larger. The slope of this function is undefined at $x = 0$, or we say that this function is not differentiable at $x = 0$.

3-5 A First Application: Marginal Analysis

7. The profit in thousands of dollars on a book sold by Books.com is given by $f(p) = -p^2 + 25p - 144$, where p is the price charged per book. The marginal profit is given by $\dfrac{df(p)}{dp} = -2p + 25$. When this derivative is evaluated at $p = \$10$, we get $\$5$ thousand dollars. Plot the profit function and the marginal profit function for p between $\$10$ and $\$15$ and comment on the graphs.

Solution: Use column A for incremental values of p and column B to evaluate the function. Then use the Chart Wizard to obtain a plot of profit versus the price per book. From the graph of $f(p) = -p^2 + 25p - 144$, we see that the maximum profit occurs when $p = \$12.50$ per book:

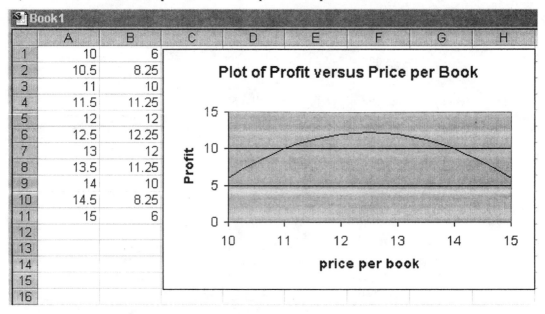

From the graph of $\dfrac{df(p)}{dp} = -2p + 25$, we see that the marginal profit function is equal to zero when $p = \$12.50$:

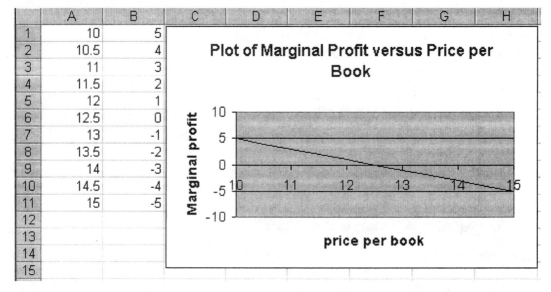

3-6 Limits and Continuity: Numerical & Graphical Approaches

Evaluating Limits Numerically

8. Evaluate the limit $\lim\limits_{x \to 0} \dfrac{x-1}{x+1}$.

Solution: In the following spreadsheet, enter the formula, =(A1-1)/(A1+1), into cell B1 and copy it to columns B and D. The values returned to the spreadsheet suggest that the limit as x approaches 0 is −1.

	A	B	C	D
1	-0.1	(A1-1)/(A1+1)	0.1	(C1-1)/(C1+1)
2	-0.01	(A2-1)/(A2+1)	0.01	(C2-1)/(C2+1)
3	-0.001	(A3-1)/(A3+1)	0.001	(C3-1)/(C3+1)
4	-0.0001	(A4-1)/(A4+1)	0.0001	(C4-1)/(C4+1)
5	-0.00001	(A5-1)/(A5+1)	0.00001	(C5-1)/(C5+1)

	A	B	C	D
1	-0.1	-1.222222222	0.1	-0.81818182
2	-0.01	-1.02020202	0.01	-0.98019802
3	-0.001	-1.002002002	0.001	-0.998002
4	-0.0001	-1.00020002	0.0001	-0.99980002
5	-0.00001	-1.00002	0.00001	-0.99998

9. Evaluate $\lim\limits_{x \to 0}(1+x)^{1/x}$.

Solution: In the following spreadsheet, enter the formula, =(1+A1)^(1/A1), into cell B1 and copy it to columns B and D. The table suggests that the limit as x approaches 0 exists. The constant, e, which the limit approaches, is to 12 significant digits 2.71828182846. More will be said later about e.

	A	B	C	D
1	-0.1	(1+A1)^(1/A1)	0.1	(1+C1)^(1/C1)
2	-0.01	(1+A2)^(1/A2)	0.01	(1+C2)^(1/C2)
3	-0.001	(1+A3)^(1/A3)	0.001	(1+C3)^(1/C3)
4	-0.0001	(1+A4)^(1/A4)	0.0001	(1+C4)^(1/C4)
5	-0.00001	(1+A5)^(1/A5)	0.00001	(1+C5)^(1/C5)

	A	B	C	D
1	-0.1	=(1+A1)^(1/A1)		2.59374246
2	-0.01	2.731999026	0.01	2.704813829
3	-0.001	2.719642216	0.001	2.716923932
4	-0.0001	2.718417755	0.0001	2.718145927
5	-0.00001	2.71829542	0.00001	2.718268237

Estimating Limits Graphically

10. Investigate the limit $\lim\limits_{x \to 0} \dfrac{x-1}{x+1}$ graphically.

Solution: The following plot indicates that $f(x)$ is close to −1 when x is near 0.

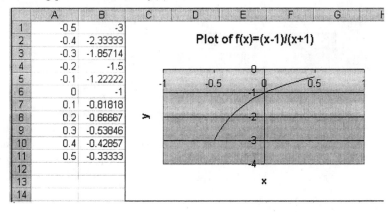

	A	B
1	-0.5	-3
2	-0.4	-2.33333
3	-0.3	-1.85714
4	-0.2	-1.5
5	-0.1	-1.22222
6	0	-1
7	0.1	-0.81818
8	0.2	-0.66667
9	0.3	-0.53846
10	0.4	-0.42857
11	0.5	-0.33333
12		
13		
14		

Plot of f(x)=(x-1)/(x+1)

11. Investigate the limit $\lim_{x \to 0}(1+x)^{1/x}$ graphically.

Solution: The following plot shows the behavior of this rather peculiar function. The function is not defined when $x = 0$ because the exponent $1/x$ is not defined when $x = 0$. However, it appears that the function is continuous otherwise and approaches $e = 2.71828182846$ as x approaches 0.

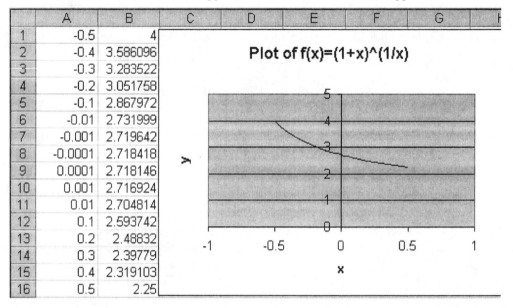

	A	B
1	-0.5	4
2	-0.4	3.586096
3	-0.3	3.283522
4	-0.2	3.051758
5	-0.1	2.867972
6	-0.01	2.731999
7	-0.001	2.719642
8	-0.0001	2.718418
9	0.0001	2.718146
10	0.001	2.716924
11	0.01	2.704814
12	0.1	2.593742
13	0.2	2.48832
14	0.3	2.39779
15	0.4	2.319103
16	0.5	2.25

Plot of f(x)=(1+x)^(1/x)

3-7 Limits and Continuity: Algebraic Approach

Non-Closed Form Functions

12. Discuss the one-sided limits at $x = 1$ and $x = 2$ for the function, $f(x) = \begin{cases} x & \text{if } x < 1 \\ x^2 & \text{if } 1 \le x \le 2 \\ x^3 & \text{if } x > 2 \end{cases}$.

Solution: Consider the Excel spreadsheet derived from the function:

	A	B	C	D
1	x	f(x)	x	f(x)
2	0.9	0.9	1.9	3.61
3	0.99	0.99	1.99	3.9601
4	0.999	0.999	1.999	3.996001
5	0.9999	0.9999	1.9999	3.9996
6	1.0001	1.0002	2.0001	8.0012
7	1.001	1.002001	2.001	8.012006
8	1.01	1.0201	2.01	8.120601
9	1.1	1.21	2.1	9.261

The following one-sided limits are observed:

$$\lim_{x \to 1^-} f(x) = 1, \ \lim_{x \to 1^+} f(x) = 1, \ \lim_{x \to 2^-} f(x) = 4, \text{and } \lim_{x \to 2^+} f(x) = 8 \, .$$

The function is seen to be continuous at $x = 1$ but discontinuous at $x = 2$.

Limits at Infinity

13. Examine the behavior of the following function as x approaches plus infinity and as x approaches minus infinity.

$$f(x) = \frac{2x^3 - x + 12}{4x^3 - x^2 + 6}$$

Solution: The formula, =(2*A2^3-A2+12)/(4*A2^3-A2^2+6), is entered into cell B2 and a similar expression is entered into cell D2. A click-and-drag produces the output shown below:

	A	B	C	D
1	x	f(x)	x	f(x)
2	-10	0.483146	10	0.512545
3	-100	0.498726	100	0.50123
4	-1000	0.499875	1000	0.500125
5	-10000	0.499987	10000	0.500012
6	-100000	0.499999	100000	0.500001

From the Excel output, we see that $\lim_{x \to -\infty} f(x) = 0.5$ and $\lim_{x \to +\infty} f(x) = 0.5$.

Chapter Review Exercises

1. The number of CD units shipped (in millions) is shown in the following table for the years 1995 through 1999. Find the average rate of change from 1995 to 1996, 1997, 1998, and 1999. Also find the average rate of change from year to year.

Year	CD Units
1995	723
1996	779
1997	759
1998	847
1999	939

2. Use Excel to construct a graph for a company that experienced sales equal to $300 million for the year 1995 and then experienced yearly rates of change equal to $25 million for 1996, $30 million for 1997, $30 million for 1998, $10 million for 1999, and $50 million for 2000.

3. The air temperature one spring morning was given by the function $f(t) = 50 + 2t^2$, t hours after 7:00. Find the average rate of change from 2 to $2 + h$, when $h = 1, 0.1, 0.01, 0.001$, and 0.0001.

4. Find the derivative of the common log function when $x = 1$ numerically.

5. Graph the function $f(x) = (x - 1)^3$ for x between -2 and 2. Evaluate the slope of the secant line connecting the points $(-h, f(-h))$ and $(h, f(h))$ for $h = 1, 0.1, 0.01, 0.001$, and 0.0001. What is the slope of the tangent line to $f(x)$ at the point $(0, -1)$? What is the connection between the slopes of the secant lines and the slope of the tangent line at $(0, -1)$?

6. Find the derivative of the function $f(x) = \sqrt{x}$. Plot the graph of the function and investigate the behavior of $\dfrac{f(h) - f(-h)}{2h}$ as h approaches 0.

7. A company's profit per day from producing x units of a new product is given by the function, $P(x) = -\dfrac{x^2}{20000} + 2.44x - 5000$, for values of x between 0 and $50,000$. Graph $P(x)$ for values of x between $20,000$ and $30,000$. Find the derivative of $P(x)$ with respect to x. For what value of x is the derivative equal to 0? What is the maximum value of $P(x)$ and for what value of x does this maximum occur?

8. Evaluate the following limit numerically and graphically: $\lim\limits_{x \to +\infty} \left(1 + \dfrac{1}{x}\right)^x$.

9. Discuss the one-sided limits at $x = 0$ and $x = 5$ for $f(x) = \begin{cases} 1 & \text{if } x < 0 \\ x & \text{if } 0 \le x \le 5 \\ \sqrt{x} & \text{if } x > 5 \end{cases}$.

CHAPTER 4

Techniques of Differentiation

4-1 The Product and Quotient Rules

1. The derivative of the function $f(x) = (3x - 2x^2)(5 + 4x)$ is found to be $\dfrac{dy}{dx} = 15 + 4x - 24x^2$ by use of

the product rule. The slope of the tangent line to $f(x)$ at $x = 2$ is equal to the derivative when it is

evaluated at $x = 2$ or $15 + 4(2) - 24(2)^2 = -73$. Verify this by evaluating $\displaystyle\lim_{h \to 0} \dfrac{f(2+h) - f(2-h)}{2h}$.

Solution: Enter the values of h into column A. The required formulas to obtain $f(2 + h), f(2 - h)$, and

$\dfrac{f(2+h) - f(2-h)}{2h}$ are respectively shown in cells B1, C1, and D1. Click-and-drag operations produce

the results shown in cells A2:D5 of the spreadsheet:

	A	B	C	D
1	1	=(3*(2+A1)-2*(2+A1)^2)*(5+4*(2+A1))	=(3*(2-A1)-2*(2-A1)^2)*(5+4*(2-A1))	=(B1-C1)/(2*A1)
2	0.1	-33.768	-19.152	-73.08
3	0.01	-26.734608	-25.274592	-73.0008
4	0.001	-26.07304601	-25.92704599	-73.000008
5	0.0001	-26.00730046	-25.99270046	-73.00000008

As h decreases, Excel returns values to column D of the spreadsheet that approach -73. Therefore,

$\displaystyle\lim_{h \to 0} \dfrac{f(2+h) - f(2-h)}{2h} = -73.$

2. The derivative of the function, $f(x) = \dfrac{x-1}{2x+3}$, is found to be $\dfrac{dy}{dx} = \dfrac{5}{(2x+3)^2}$ by use of the quotient

rule. The slope of the tangent line at $x = 0$ is found by evaluating the expression for the derivative at

$x = 0$. This is seen to be $\dfrac{5}{9} = 0.55555\ldots$. Verify this by evaluating $\displaystyle\lim_{h \to 0} \dfrac{f(0+h) - f(0-h)}{2h}$.

Solution: The values of h are shown in column A, $f(0 + h)$ is shown in column B, $f(0 - h)$ is shown in

column C and $\dfrac{f(0+h) - f(0-h)}{2h}$ is shown in column D of the Excel spreadsheet:

	A	B	C	D
1	1	=(A1-1)/(2*A1+3)	=(-A1-1)/(2*-A1+3)	=(B1-C1)/(2*A1)
2	0.1	-0.28125	-0.392857143	0.558035714
3	0.01	-0.32781457	-0.338926174	0.555580248
4	0.001	-0.332778148	-0.33388926	0.555555802
5	0.0001	-0.333277781	-0.333388893	0.555555558

As h decreases, the values returned to column D of the spreadsheet approach 0.55555... . Therefore, $\lim\limits_{h\to 0}\dfrac{f(0+h)-f(0-h)}{2h}=0.55555...$.

4-2 The Chain Rule

3. If \$1,000 is invested at x % convertible quarterly for 10 years, the amount at the end of 10 years is given by $f(x)=1000\left(1+\dfrac{x}{400}\right)^{40}$. Applying the chain rule, the derivative is $\dfrac{dy}{dx}=100\left(1+\dfrac{x}{400}\right)^{39}$. If the interest rate is 5% convertible quarterly, the instantaneous rate of change in the amount function is equal to $100\left(1+\dfrac{5}{400}\right)^{39}=\$162.33$. Compare this value to average change in the amount function if the interest rate were increased from 5% to 6%, 5.5%, 5.1%, 5.01% and 5.001%.

Solution: The Excel computations and results are shown in the next two panels.

	A	B	C	D
1	Quarterly %, x	Amount at x% Quarterly	Amount at 5%	Average Change
2	6	=1000*(1+A2/400)^40	=1000*(1+5/400)^40	=(B2-C2)/(A2-5)
3	5.5	=1000*(1+A3/400)^40	=1000*(1+5/400)^40	=(B3-C3)/(A3-5)
4	5.1	=1000*(1+A4/400)^40	=1000*(1+5/400)^40	=(B4-C4)/(A4-5)
5	5.01	=1000*(1+A5/400)^40	=1000*(1+5/400)^40	=(B5-C5)/(A5-5)
6	5.001	=1000*(1+A6/400)^40	=1000*(1+5/400)^40	=(B6-C6)/(A6-5)

	A	B	C	D
1	Quarterly %, x	Amount at x% Quarterly	Amount at 5%	Average Change
2	6	1814.018409	1643.619463	170.3989452
3	5.5	1726.770771	1643.619463	166.3026151
4	5.1	1659.931147	1643.619463	163.1168389
5	5.01	1645.243573	1643.619463	162.4109712
6	5.001	1643.781804	1643.619463	162.3406028

Note that the average change in the amount function when 6% quarterly is compared to 5% quarterly is equal to the gain in the amount when the quarterly rate is changed by one percentage point, and this is approximately equal to the derivative evaluated when $x=5\%$. Thus, one possible interpretation of the derivative evaluated at 5% is $\dfrac{dy}{dx}=100\left(1+\dfrac{x}{400}\right)^{39}$ evaluated at 5% is approximately the change in the amount after 10 years due to changing x from 5% to 6%. The derivative evaluated at 8% is equal to $100\left(1+\dfrac{8}{400}\right)^{39}=\$216.47$. Therefore, we could state that for this investment, increasing the rate from 8% to 9% would increase the amount earned by approximately \$216.47. The actual increase is \$227.15.

For the function $f(x) = 1000\left(1 + \dfrac{x}{400}\right)^{40}$, the following Excel spreadsheet gives the values of x in column A, $f(x + 1)$ in column B, $f(x)$ in column C, $f(x + 1) - f(x)$ in column D, and the derivative of $f(x)$ in column E:

	A	B	C	D	E
1	x	f(x+1)	f(x)	f(x+1)-f(x)	Derivative of f(x)
2	0	1105.033	1000	105.033	100
3	1	1220.794	1105.033	115.7612	110.2277317
4	2	1348.349	1220.794	127.5544	121.4720633
5	3	1488.864	1348.349	140.5151	133.8311278
6	4	1643.619	1488.864	154.7557	147.4122509
7	5	1814.018	1643.619	170.3989	162.3327865
8	6	2001.597	1814.018	187.5789	178.7210255
9	7	2208.04	2001.597	206.4423	196.7171836
10	8	2435.189	2208.04	227.1493	216.4744768
11	9	2685.064	2435.189	249.8749	238.16029
12	10	2959.874	2685.064	274.8101	261.9574476

Note that the values in column E are reasonable approximations to those in column D.

4. If the function $f(x) = (3x - 2x^2)^3$ is differentiated using the chain rule, the result is $(9 - 12x)(3x - 2x^2)^2$. Compare the value of the derivative with $f(x + 1) - f(x)$ for $x = 0, 1, \cdots, 10$.

Solution: Enter the numbers 0 through 10 are into cells A2 through A12. Then enter the function, =(3*(A2+1)-2*(A2+1)^2)^3, is into B2; the function, =(3*(A2)-2*(A2)^2)^3, into cell C2; the function, =B2-C2, into cell D2; and the function, =(9-2*A2)*(3*A2-2*(A2)^2)^2, into cell E2. Perform click-and-drags to produce the following output:

	A	B	C	D	E
1	x	f(x+1)	f(x)	f(x+1)-f(x)	Derivative of f(x)
2	0	1	0	1	0
3	1	-8	1	-9	-3
4	2	-729	-8	-721	-60
5	3	-8000	-729	-7271	-2187
6	4	-42875	-8000	-34875	-15600
7	5	-157464	-42875	-114589	-62475
8	6	-456533	-157464	-299069	-183708
9	7	-1124864	-456533	-668331	-444675
10	8	-2460375	-1124864	-1335511	-940992
11	9	-4913000	-2460375	-2452625	-1804275
12	10	-9129329	-4913000	-4216329	-3207900
13					

In this example, the derivative of $f(x)$ is not a good approximation of $f(x + 1) - f(x)$ as was the case in example 3 of this section. The function, $f(x) = 1000\left(1 + \dfrac{x}{400}\right)^{40}$, is much closer to a linear function than is the function $f(x) = (3x - 2x^2)^3$. For a linear function, $\dfrac{dy}{dx} = f(x+1) - f(x)$.

4-3 Derivatives of Logarithmic and Exponential functions

Derivatives of Logarithmic Functions

5. The common logarithm function is $f(x) = \log_{10}x$ and is usually written as $f(x) = \log x$. The derivative of this function is $\dfrac{dy}{dx} = \dfrac{1}{x\ln(10)}$. Plot the common logarithm function and its derivative on the same graph and discuss the two graphs.

Solution: Enter the formula, =LOG10(A2), into cell B2 and the formula, =1/(LN(10)*A2), into cell C2. Click-and-drags produce the values shown in columns B and C of the spreadsheet:

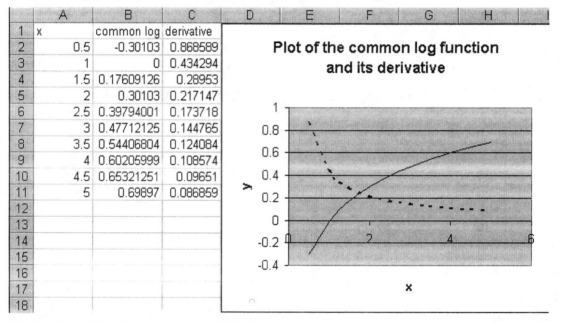

	A	B	C	D	E	F	G	H	I
1	x	common log	derivative						
2	0.5	-0.30103	0.868589						
3	1	0	0.434294						
4	1.5	0.17609126	0.28953						
5	2	0.30103	0.217147						
6	2.5	0.39794001	0.173718						
7	3	0.47712125	0.144765						
8	3.5	0.54406804	0.124084						
9	4	0.60205999	0.108574						
10	4.5	0.65321251	0.09651						
11	5	0.69897	0.086859						
12									
13									
14									
15									
16									
17									
18									

Use the Chart Wizard's XY scatter to build the graph with data points connected by smoothed lines without markers. The solid curve is the graph of the common log function and the dashed curve is the graph of the derivative of the common log function. When x is close to 0, the graph of the common log is rising steeply and the derivative is a large positive number. As x increases, the graph of the common log becomes less steep, although it continues to rise. The derivative reflects the behavior of the slope of the tangent line to the curve.

Derivatives of Exponential Functions

6. The derivative of the exponential function, $f(x) = e^x$, is $\dfrac{dy}{dx} = e^x$. Illustrate this by choosing two

arbitrary values say $x = -2$ and $x = 2$ and considering the behavior of $\dfrac{f(x+h)-f(x-h)}{2h}$ as h

approaches 0.

Solution: Enter the following formulas into the cells of the Excel spreadsheet: =EXP(2+A2) in cell B2, =EXP(2-A2) in cell C2, =(B2-C2)/(2*A2) in cell D2, and =EXP(2) in cell E2. Click-and-drags produced the results shown in cells B2 through E6. Similarly, enter =EXP(-2+A8) into cell B8, =EXP(-2-A8) into cell C8, =(B8-C8)/(2*A8) into cell D8, and =EXP(-2) into cell E8. Click-and-drags produce the results shown in cells B8 through E12:

	A	B	C	D	E
1	h	exp(2+h)	exp(2-h)	(exp(2+h)-exp(2-h))/2h	exp(2)
2	1	20.0855369	2.718282	8.683627547	7.389056
3	0.1	8.16616991	6.685894	7.401377351	7.389056
4	0.01	7.46331735	7.315534	7.38917925	7.389056
5	0.001	7.39644885	7.381671	7.38905733	7.389056
6	0.0001	7.38979504	7.388317	7.389056111	7.389056
7	h	exp(-2+h)	exp(-2-h)	(exp(-2+h)-exp(-2-h))/2h	exp(-2)
8	1	0.36787944	0.049787	0.159046186	0.135335
9	0.1	0.14956862	0.122456	0.135560955	0.135335
10	0.01	0.13669543	0.133989	0.135337539	0.135335
11	0.001	0.13547069	0.1352	0.135335306	0.135335
12	0.0001	0.13534882	0.135322	0.135335283	0.135335

The results in the spreadsheet illustrate that $\lim\limits_{h\to 0} \dfrac{f(x+h)-f(x-h)}{2h} = f(x)$ when $f(x) = e^x$. The limiting

values in the table are equal to e^{-2} and e^2.

Applications

7. The population of a bacterial culture is given by the following logistics growth function, where y is the number of bacteria and x is the time in days:

$$y = f(x) = \frac{1150}{1 + e^{-0.2x}}$$

The slope of this growth curve is found by using the quotient rule to be $\dfrac{dy}{dx} = \dfrac{230 - e^{-0.2x}}{\left(1 + e^{-0.2x}\right)^2}$.

a) Graph both the logistics curve and its derivative and comment on both.
b) What is the limit of $f(x)$ as x approaches infinity?

Solution: Enter the number of days into column A. Then enter the formula, =1150/(1+EXP(-0.2*A2)), into cell B2 and perform a click-and-drag. Finally, use the Chart Wizard to construct the following plot:

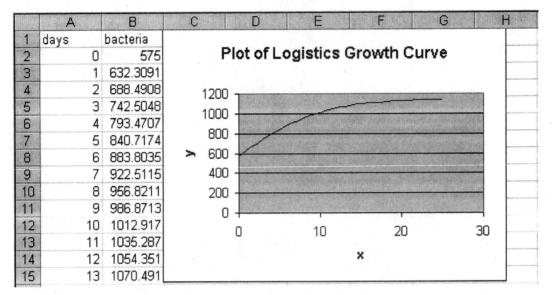

	A	B
1	days	bacteria
2	0	575
3	1	632.3091
4	2	688.4908
5	3	742.5048
6	4	793.4707
7	5	840.7174
8	6	883.8035
9	7	922.5115
10	8	956.8211
11	9	986.8713
12	10	1012.917
13	11	1035.287
14	12	1054.351
15	13	1070.491

To plot the derivative of the function, enter the values of x from 0 to 25 in column A, and the formula, =230*EXP(-0.2*A2)/(1+EXP(-0.2*A2))^2, is into cell B2. Then perform a click-and-drag to obtain the values in column B:

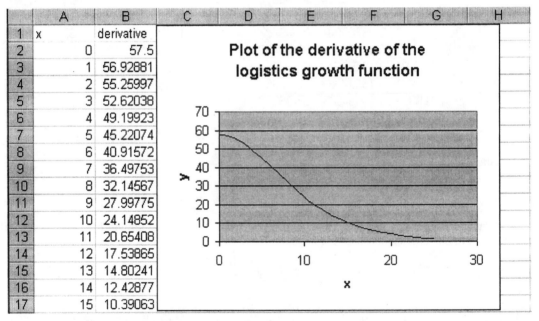

	A	B
1	x	derivative
2	0	57.5
3	1	56.92881
4	2	55.25997
5	3	52.62038
6	4	49.19923
7	5	45.22074
8	6	40.91572
9	7	36.49753
10	8	32.14567
11	9	27.99775
12	10	24.14852
13	11	20.65408
14	12	17.53865
15	13	14.80241
16	14	12.42877
17	15	10.39063

The derivative plot confirms what we see in the plot of the growth function. The slope of the logistics growth curve starts out steeply and decreases and finally approaches zero. This tells us that as time goes by, the growth curve becomes very nearly horizontal. In many growth situations, there is an upper limit past which growth may not occur. The logistics growth curve is a very good model in many such cases. For part b) of the problem, the next panel considers the limit as x approaches infinity. The number of bacteria levels off at 1150 and remains constant.

	A	B
1	days	bacteria
2	25	1142.303
3	50	1149.948
4	75	1150
5	100	1150

4-4 Derivatives of Trigonometric Functions

8. The derivative of the tangent function is the secant function squared. But since the secant function is the reciprocal of the cosine function, it is also true that the derivative of the tangent function is the reciprocal of the cosine function squared. If $f(x) = \tan x$, then we have:

$$\frac{dy}{dx} = \frac{1}{\cos^2 x}$$

a) Plot both the tangent function and its derivative for x between 0 and $\pi/2$ and discuss the graphs.
b) Also look at the behavior of the tangent function as x approaches $\pi/2$ from the left.

Solution: For part a) of the problem, enter the values between 0 and 1.5 into column A. Then enter the formula, =TAN(A2) into cell B2. Perform a click-and-drag to obtain the values in column B. Finally, use the Chart Wizard to plot the tangent function as shown in the following panel:

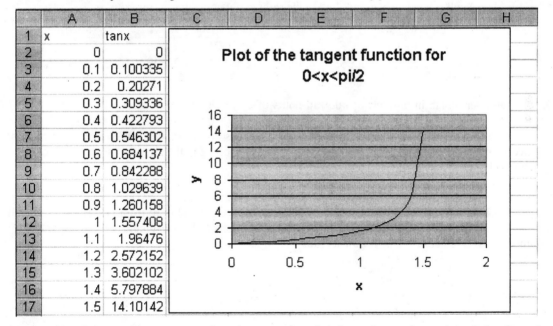

For part b) of the problem, enter values between 0 and 1.5 are into column A and the formula, =1/(COS(A2))^2, into cell B2. Then perform a click-and-drag to obtain values in column B. Finally, use the Chart Wizard to plot the derivative of the tangent function as shown in the following panel.

	A	B	C	D	E	F	G	H
1	x	derivative of tanx						
2	0	1						
3	0.1	1.010067						
4	0.2	1.041091						
5	0.3	1.095689						
6	0.4	1.178754						
7	0.5	1.298446						
8	0.6	1.468043						
9	0.7	1.70945						
10	0.8	2.060156						
11	0.9	2.587999						
12	1	3.425519						
13	1.1	4.860281						
14	1.2	7.615964						
15	1.3	13.97514						
16	1.4	34.61546						
17	1.5	199.85						
18								

Plot of the derivative of the Tangent function

Note that as x approaches $\pi/2 = 1.570796327$, the slope of the tangent line grows larger and larger. The following panel shows the behavior of the tangent function as x gets close to $\pi/2$. Now enter the formula, =TAN(B1), into cell B2 and a click-and-drag to the right to obtain values in row 2:

	A	B	C	D	E	F	G	H
1	x	1.51	1.52	1.53	1.54	1.55	1.56	1.57
2	tanx	16.42809	19.66953	24.49841	32.46114	48.07848	92.6205	1255.766

9. The sales per month in millions of kilowatt hours of electricity in Midwest city from January 1997 through January 2000 is given by the equation $y = f(x) = 30 + 0.5x + 12\cos(\pi x/3)$. In the equation $x = 1$ corresponds to January 1997 and $x = 36$ corresponds to December 1999. Plot the function. Investigate the behavior of $f(x)$ close to June of 1997 by looking at the slopes of secant lines close to June of 1997.

Solution: Enter the numbers 1 through 36 into cells A2 through A37. Then enter the formula, =30+0.5*A2+12*COS(1.0472*A2), into cell B2. Execute a click-and-drag from cell B2 through cell B37 to obtain the sales per month of electricity. Use the Chart Wizard together with the values in columns A and B to construct the following plot:

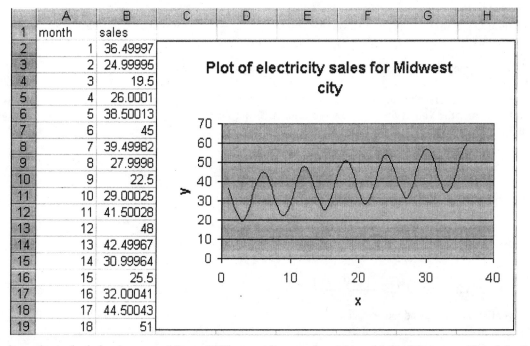

	A	B	C	D	E	F	G	H
1	month	sales						
2	1	36.49997						
3	2	24.99995						
4	3	19.5						
5	4	26.0001						
6	5	38.50013						
7	6	45						
8	7	39.49982						
9	8	27.9998						
10	9	22.5						
11	10	29.00025						
12	11	41.50028						
13	12	48						
14	13	42.49967						
15	14	30.99964						
16	15	25.5						
17	16	32.00041						
18	17	44.50043						
19	18	51						

To investigate the behavior around June, 1997 secant lines to the right and left of 6 are considered. In the following panel, enter the formulas, =30+0.5*(6-A2)+12*COS(1.0472*(6-A2)) into cell B2, =45 into cell C2, =30+0.5*(6+A2)+12*COS(1.0472*(6+A2)) into cell D2, =(C2-B2)/A2 into cell E2, and =(D2-C2)/A2 into cell F2. Then execute a click-and-drag to obtain the following spreadsheet:

	A	B	C	D	E	F
1	h	f(6-h)	f(6)	f(6+h)	(f(6)-f(6-h))/h	(f(6+h)-f(6))/h
2	1	38.50013	45	39.49982	6.49987276	-5.50017814
3	0.1	44.88428	45	44.98424	1.15719133	-0.15755993
4	0.01	44.99434	45	45.00434	0.56561244	0.4340183
5	0.001	44.99949	45	45.00049	0.50639513	0.4932356
6	0.0001	44.99995	45	45.00005	0.50047334	0.49915739

The slopes of the secant lines on the left of $x = 6$ are positive, indicating that the electricity usage is increasing as June is approached. The slope of the secant line to the right of $x = 6$ for $h = 0.1$ is negative, indicating that the electricity usage has begun to decrease shortly after June.

4-5 Implicit Differentiation

10. The circle with center at (0, 0) and having radius 1 is defined by the equation $x^2 + y^2 = 1$. Find the slope of the tangent line to the circle at the point $\left(\dfrac{\sqrt{2}}{2}, \dfrac{\sqrt{2}}{2} \right)$ by using both implicit differentiation and by using the equation of the upper half of the circle, coupled with the definition of the derivative $\lim_{h \to 0} \dfrac{f(x+h) - f(x-h)}{2h}$ where $f(x) = \sqrt{1 - x^2}$. Show that the same answer is obtained.

Solution: Using implicit differentiation, we obtain $2x + 2y\dfrac{dy}{dx} = 0$. Solving for the derivative, we obtain

$\dfrac{dy}{dx} = -\dfrac{x}{y}$. Using the expression obtained for the derivative, the slope of the tangent line at $\left(\dfrac{\sqrt{2}}{2}, \dfrac{\sqrt{2}}{2}\right) \approx$

(0.707, 0.707) is -1. We can use Excel, coupled with the definition of the derivative, to solve the same problem. Enter the values for h are into column A. Enter the formulas, =SQRT(1-(0.707-A2)^2), into cell B2, =SQRT(1-(0.707+A2)^2) into cell C2, and =(C2-B2)/(2*A2) into cell D2. Then execute a click-and-drag to obtain the following spreadsheet:

	A	B	C	D
1	h	f(0.707-h)	f(0.707+h)	(f(0.707+h)-f(0.707-h))/2h
2	0.1	0.794702	0.590551	-1.020751968
3	0.01	0.717071	0.697073	-0.999897964
4	0.001	0.708212	0.706212	-0.999700044
5	0.0001	0.707314	0.707114	-0.999698066

The output indicates that $\displaystyle\lim_{h\to 0}\dfrac{f(0.707+h)-f(0.707-h)}{2h} = -1$, the same as obtained by using implicit differentiation. As an added exercise, substitute SQRT(2)/2 in your formulas rather than the 0.707 approximation to see what impact that the rounded number has on the values that Excel returns.

Chapter Review Exercises

1. Find the derivative of $f(x) = (x^3 - x^2)(x + x^4)$ by using the product rule. Evaluate this derivative for $x = 2$. Confirm your answer by using the definition of the derivative.

2. Find the derivative of $f(x) = \dfrac{x}{x^3 + 4}$ by using the quotient rule. Evaluate this derivative for $x = 1$. Confirm your answer by using the definition of the derivative.

3. Find the derivative of $f(x) = (1 + x)^4$ by use of the chain rule. Evaluate this derivative for $x = 2$. Confirm your answer by using the definition of the derivative.

4. Find the derivative of $f(x) = \ln(x^2 + 4)$ and evaluate the derivative for $x = 3$. Confirm your answer by using the definition of the derivative.

5. Graph the function, $f(x) = e^{x^2}$, find its derivative, and evaluate it for $x = 0$. Confirm your answer by using the definition of the derivative.

6. Plot the logistics curve, $f(x) = \dfrac{10}{1 + e^{-x}}$. Evaluate the derivative when $x = 10$. What is the limit of $f(x)$ as x approaches infinity?

7. Plot both the cotangent function and its derivative for x between 0 and $\pi/2$ and discuss the graphs. Also look at the behavior of the cotangent function as x approaches 0 from the right.

8. Plot the function, $f(x) = 0.5x + 0.05 \sin \pi x$, for x between 0 and 3. Find the derivative when $x = 1.5$ using the rules of differentiation and by using the basic definition.

9. Find the slope of the tangent line to the ellipse $x^2 + 4y^2 = 4$ at the point $(\sqrt{2}, -\frac{\sqrt{2}}{2})$ in two ways. First, find the slope by using implicit differentiation. Second, solve the equation $x^2 + 4y^2 = 4$ for the equation of the lower half of the ellipse and use that equation and the basic definition of the derivative to find the slope of the tangent line at the point $(\sqrt{2}, -\frac{\sqrt{2}}{2})$.

CHAPTER 5

Applications of the Derivative

5-1 Maxima and Minima

1. Find the relative and absolute maxima and minima of $f(x) = 3x^2 - 5x + 7$ on the interval $[-1, 5]$. Use graphical as well as calculus techniques.

Solution: Enter the numbers -1, -0.9, $\cdots$, 5 into column A and the formula, $=3*A1\wedge2-5*A1+7$, into cell B1. A click-and-drag returns 61 points on the graph of $f(x)$. Now use the Chart Wizard to construct the following graph:

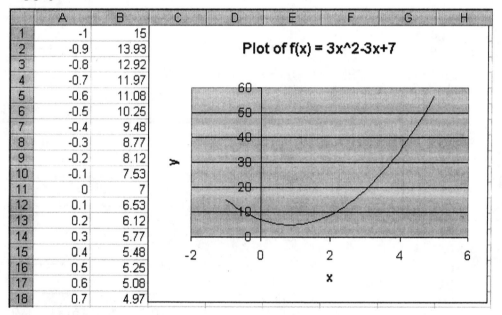

The endpoints are seen to be $(-1, 15)$ and $(5, 57)$. It is also clear from the graph that a stationary point is likely somewhere between 0 and 2.

The following panels show how to zoom in on the potential stationary point. In columns A and B we see that the stationary point is between $(0.8, 4.92)$ and $(0.9, 4.93)$ because the curve changes direction here. Therefore, we can reduce the range in the values being plotted for both x and y.

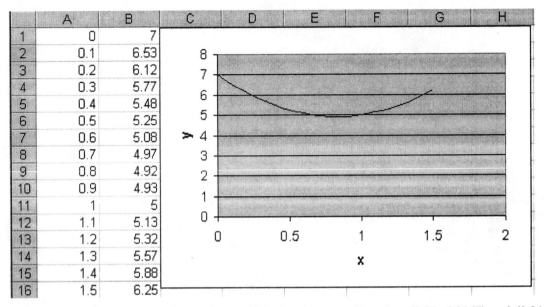

	A	B	C	D	E	F	G	H
1	0	7						
2	0.1	6.53						
3	0.2	6.12						
4	0.3	5.77						
5	0.4	5.48						
6	0.5	5.25						
7	0.6	5.08						
8	0.7	4.97						
9	0.8	4.92						
10	0.9	4.93						
11	1	5						
12	1.1	5.13						
13	1.2	5.32						
14	1.3	5.57						
15	1.4	5.88						
16	1.5	6.25						

The next panel shows that the curve changes direction between the points (0.83, 4.9167) and (0.84, 4.9168). We have reduced the range in the values being plotted for both x and y again.

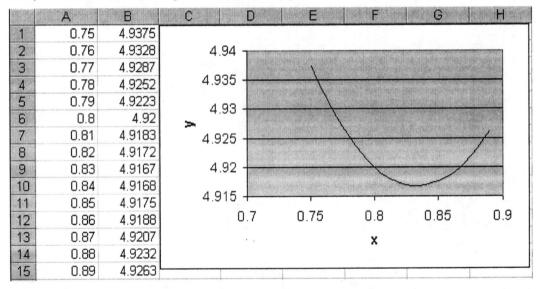

	A	B	C	D	E	F	G	H
1	0.75	4.9375						
2	0.76	4.9328						
3	0.77	4.9287						
4	0.78	4.9252						
5	0.79	4.9223						
6	0.8	4.92						
7	0.81	4.9183						
8	0.82	4.9172						
9	0.83	4.9167						
10	0.84	4.9168						
11	0.85	4.9175						
12	0.86	4.9188						
13	0.87	4.9207						
14	0.88	4.9232						
15	0.89	4.9263						

Using yet another range-reduction procedure, the next panel shows that the curve changes direction between the points (0.833, 4.916667) and (0.834, 4.916668).

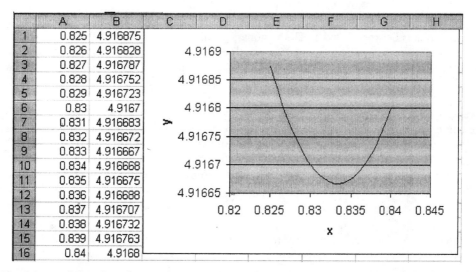

	A	B	C	D	E	F	G	H
1	0.825	4.916875						
2	0.826	4.916828						
3	0.827	4.916787						
4	0.828	4.916752						
5	0.829	4.916723						
6	0.83	4.9167						
7	0.831	4.916683						
8	0.832	4.916672						
9	0.833	4.916667						
10	0.834	4.916668						
11	0.835	4.916675						
12	0.836	4.916688						
13	0.837	4.916707						
14	0.838	4.916732						
15	0.839	4.916763						
16	0.84	4.9168						

To verify that we have found a stationary point, we now turn to our knowledge of calculus. The derivative of $f(x)$ is $\dfrac{dy}{dx} = 6x - 5$. Setting this equal to 0 and solving for x we find $x = 5/6 = 0.833$, the same value we found graphically. Summarizing, we have found that there is

- a relative maximum at the point $(-1, 15)$
- a relative minimum at the point $(0.833, 4.917)$, which is also the absolute minimum on $[-1, 5]$
- a relative maximum at the point $(5, 57)$, which is also the absolute maximum on $[-1, 5]$.

2. Find the relative and absolute maxima and minima of $f(x) = x^3 - 1.5x^2 + 0.5x$ that lie within the interval $[-0.5, 1.5]$. Use graphical as well as calculus techniques.

Solution: Enter $-0.5, -0.4, \cdots , 1.5$ into column A and the formula, =A1^3-1.5*A1^2+0.5*A1, into cell B1. A click-and-drag returns 21 points on the graph of $f(x)$. Use the Chart Wizard to build the following graph:

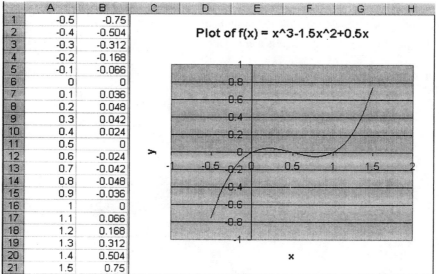

	A	B	C	D	E	F	G	H
1	-0.5	-0.75						
2	-0.4	-0.504						
3	-0.3	-0.312						
4	-0.2	-0.168						
5	-0.1	-0.066						
6	0	0						
7	0.1	0.036						
8	0.2	0.048						
9	0.3	0.042						
10	0.4	0.024						
11	0.5	0						
12	0.6	-0.024						
13	0.7	-0.042						
14	0.8	-0.048						
15	0.9	-0.036						
16	1	0						
17	1.1	0.066						
18	1.2	0.168						
19	1.3	0.312						
20	1.4	0.504						
21	1.5	0.75						

Plot of f(x) = x^3-1.5x^2+0.5x

The endpoints are seen to be (−0.5, −0.75) and (1.5, 0.75). It is also clear from the graph that there are two stationary points between 0 and 1. The stationary points are located where the first derivative, $3x^2 - 3x + 0.5$, is equal to zero. Setting $3x^2 - 3x + 0.5$ equal to zero and using the quadratic equation formula to find the solutions, we find $x = 0.211$ and 0.789 to be critical points. From the graph, we know that $(0.211, 0.04811)$ is a relative maximum and $(0.789, -0.04811)$ is a relative minimum.

If we were to use the method of zooming in to find the relative maximum at $x = 0.211$, we would note that the curve changes direction between 0.2 and 0.3 in the function values accompanying the graph above. Then in the panel below, we see that the function values increase from $x = 0.16$ to $x = 0.21$. From $x = 0.22$ through $x = 0.25$, the function values are decreasing. The relative maximum is between $x = 0.21$ and $x = 0.22$. In columns D and E, the function increases from $x = 0.204$ to $x = 0.211$ and then decreases for $x = 0.212$ and 0.213. This tells us that the relative maximum is between $x = 0.211$ and $x = 0.212$.

	A	B	C	D	E
1	0.16	0.045696		0.204	0.048065664
2	0.17	0.046563		0.205	0.048077625
3	0.18	0.047232		0.206	0.048087816
4	0.19	0.047709		0.207	0.048096243
5	0.2	0.048		0.208	0.048102912
6	0.21	0.048111		0.209	0.048107829
7	0.22	0.048048		0.21	0.048111
8	0.23	0.047817		0.211	0.048112431
9	0.24	0.047424		0.212	0.048112128
10	0.25	0.046875		0.213	0.048110097

Similarly, the relative minimum at $x = 0.789$ could be zoomed in on. Summarizing, we have found that there is
- an absolute minimum at the point (−0.5, −0.75)
- a relative minimum at the point (0.789, −0.04811)
- a relative maximum at the point (0.211, 0.04811)
- an absolute maximum at the point (1.5, 0.75).

3. Find the relative and absolute maxima and minima of $f(x) = \sqrt[3]{(x^2 - 4)^2} = (x^2 - 4)^{2/3}$ on the interval $[-3, 3]$. Use graphical as well as calculus techniques.

Solution: Enter the numbers −3, −2.9, ⋯ , 3 into column A and the formula, =((A1^2−4)^2)^(1/3), into cell B1. Then click-and-drag forms the function values in column B. Now use the Chart Wizard to construct the following graph:

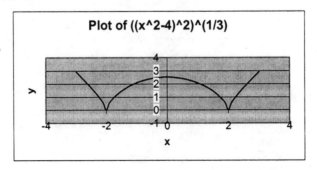

From the graph, it appears that $x = 0$ corresponds to a stationary point and that $x = -2$ and $x = 2$ correspond to singular points. The derivative of $f(x)$ is $\dfrac{dy}{dx} = \dfrac{4x}{3\sqrt[3]{(x^2 - 4)}}$. The derivative is equal to 0 when $x = 0$ and it is undefined when $x = -2$ and $x = 2$. The function $f(x)$ is defined for all values of x.

Summarizing, we have found that
- both $(-3, 2.92)$ and $(3, 2.92)$ are relative and absolute maxima on $[-3, 3]$.
- both $(-2, 0)$ and $(2, 0)$ are relative and absolute minima.
- $(0, 2.52)$ is a relative maximum on $[-3, 3]$.

5-2 Applications of Maxima and Minima

4. Customboxes.com makes boxes for industrial customers. One order has the following requirements: An open box is to be made from a square piece of cardboard, 12 inches on a side, by cutting equal squares from each corner and turning up the sides. Find the dimensions of the box with the largest volume that can be made in this manner and the volume of that box.

Solution: Let the squares that are to be cut from the corners have dimension x by x. The dimensions of the open box will be $(12 - 2x)$ by $(12 - 2x)$ by x and the volume will be $f(x) = (12 - 2x)^2 x = 144x - 48x^2 + 4x^3$. The domain of x is $0 < x < 6$. Enter the numbers $0, 0.1, 0.2, \cdots, 6$ into column A and the formula, =144*A1-48*A1^2+4*A1^3, into cell B1. Perform a click-and-drag in column B to obtain the values required for plotting. Finally, use the Chart Wizard to plot the function:

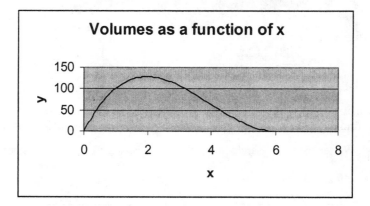

It is clear from the graph that the maximum volume occurs for x near 2. The maximum value may be found by setting the derivative equal to 0. The derivative is $\dfrac{dy}{dx} = 144 - 96x + 12x^2$. Using the quadratic formula, the solutions to the equation $144 - 96x + 12x^2 = 0$ are as follows:

$$x = \frac{96 \pm \sqrt{96^2 - 4(12)(144)}}{2(12)}$$

The solutions are $x = 2$ and $x = 6$. Since 6 is not in the domain, our only solution is $x = 2$. Therefore, the dimensions of the box are $[12 - (2)(2)]$ by $[12 - (2)(2)]$ by 2 or 8 by 8 by 2. The volume of the box is $(8)(8)(2) = 128$ cubic inches.

5. Find the positive number x that minimizes the sum of x and its reciprocal.

Solution: Mathematically, we must find the value of x for which $f(x) = x + 1/x$ is a minimum, subject to the constraint that $x > 0$. Build the graph of the function first. Enter the numbers 0.1, 0.2, $\cdots$, 3.0 into column A and he formula, =A1+1/A1, into cell B1. A click-and-drag produces the function values. Now use the Chart Wizard to construct the plot:

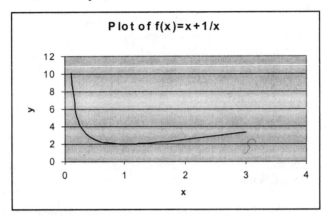

It appears that a possible minimum exists near $x = 1$. However, let's also investigate the behavior of the function when x is close to 0, and when x goes to infinity. The following panel looks at the behavior of $f(x)$ when x is close to 0 and when x is large.

	A	B	C	D	E
1	0.1	10.1		3	3.333333
2	0.01	100.01		10	10.1
3	0.001	1000.001		100	100.01
4	0.0001	10000		1000	1000.001
5	0.00001	100000		10000	10000

We note that $f(x)$ increases as x gets close to 0 and also when x becomes large. By considering the derivative, we can find the value of x that minimizes the sum of x and its reciprocal. The derivative of $f(x)$ is $\dfrac{dy}{dx} = 1 - \dfrac{1}{x^2}$. We see that $x = -1$ and $x = 1$ makes the derivative equal to 0. Since we are interested in the minimum value for $x > 0$, we choose $x = 1$. A plot of the derivative for values of x close to $x = 1$ is shown below.

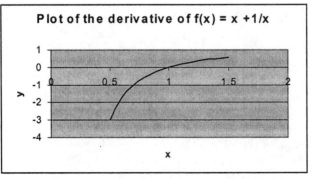

The derivative is negative to the left of $x = 1$ and positive to the right of $x = 1$. This assures us that the point $(1, 2)$ corresponds to the absolute minimum of the function.

5-3 The Second Derivative and Analyzing Graphs

Acceleration

6. If a rock is dropped from a height of 1600 feet above the ground, its height above the ground is given by the equation $f(x) = -16x^2 + 1600$ after x seconds where $0 < x < 10$. The velocity is the derivative of $f(x)$ and the acceleration is given by the second derivative of $f(x)$. Find the velocity and acceleration functions. Plot the height and the velocity functions.

Solution: The first derivative is $\dfrac{dy}{dx} = -32x$ and the second derivative is $\dfrac{d^2y}{dx^2} = -32$. Enter the time values 0, 1, $\cdots$, 10 into column A of the spreadsheet. Then enter the formula, =−16*A2^2+1600, into cell B2. A click-and-drag returns all height values. Use the Chart Wizard to plot the height function:

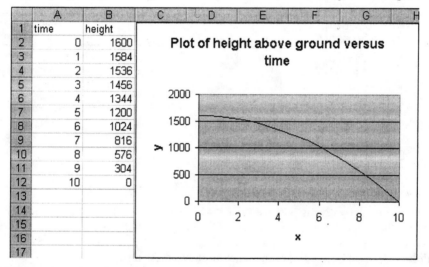

In a similar fashion, the values for velocity are found using the function, =−32A1, and plotted:

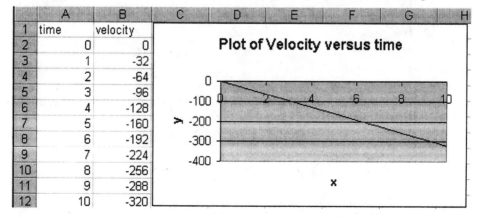

Concavity

7. Determine where the function $f(x) = \dfrac{6}{x^2 + 3}$ is concave upward and concave downward. What are the points of inflection?

Solution: The first derivative is $\dfrac{dy}{dx} = (-6)(2x)(x^2 + 3)^{-2} = \dfrac{-12x}{(x^2 + 3)^2}$ and the second derivative found

to be $\dfrac{d^2 y}{dx^2} = \dfrac{36(x^2 - 1)}{(x^2 + 3)^3}$. To plot the second derivative, enter the numbers $-2, -1.9, \cdots, 2$ into column A and the expression =(36*(A1^2-1)/(A1^2+3)^3) into cell B1 and perform a click-and-drag.

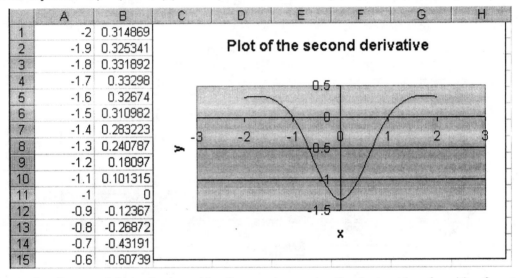

	A	B
1	-2	0.314869
2	-1.9	0.325341
3	-1.8	0.331892
4	-1.7	0.33298
5	-1.6	0.32674
6	-1.5	0.310982
7	-1.4	0.283223
8	-1.3	0.240787
9	-1.2	0.18097
10	-1.1	0.101315
11	-1	0
12	-0.9	-0.12367
13	-0.8	-0.26872
14	-0.7	-0.43191
15	-0.6	-0.60739

Plot of the second derivative

We see that the second derivative is positive for $x < -1$, negative for $-1 < x < 1$ and positive for $x > 1$. This tells us that the graph is concave upward for $x < -1$, concave downward for $-1 < x < 1$, and concave upward for $x > 1$. Since the concavity changes at $x = -1$ and $x = 1$, $(-1, 1.5)$ and $(1, 1.5)$ are points of inflection. The following plot of the function shows these results:

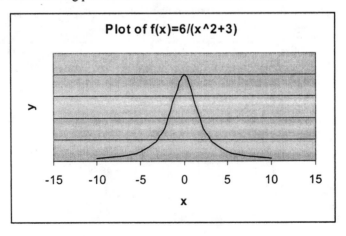

Plot of f(x)=6/(x^2+3)

8. Determine where the function $f(x) = x^4 - 5x^2 + 4$ is concave upward and concave downward. What are the points of inflection?

Solution: The first derivative is $\dfrac{dy}{dx} = 4x^3 - 10x$ and the second derivative is $\dfrac{d^2 y}{dx^2} = 12x^2 - 10$. The following is a plot of the second derivative:

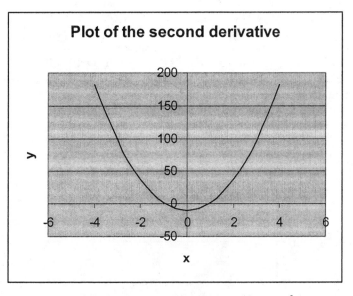

The roots of this quadratic function are found by solving the equation $12x^2 - 10 = 0$. The solutions are $x = -0.913$ and $x = 0.913$. The second derivative is positive for $x < -0.913$, negative for $-0.913 < x < 0.913$, and positive for $x > 0.913$. Therefore, the function $f(x) = x^4 - 5x^2 + 4$ is concave upward for $x < -0.913$, concave downward for $-0.913 < x < 0.913$, and concave upward for $x > 0.913$. The points of inflection are $(-0.913, 0.527)$ and $(0.913, 0.527)$. The following plot of the function shows these results.

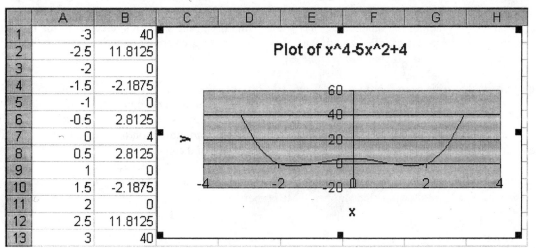

	A	B	C	D	E	F	G	H
1	-3	40						
2	-2.5	11.8125						
3	-2	0						
4	-1.5	-2.1875						
5	-1	0						
6	-0.5	2.8125						
7	0	4						
8	0.5	2.8125						
9	1	0						
10	1.5	-2.1875						
11	2	0						
12	2.5	11.8125						
13	3	40						

Analyzing Graphs

9. Analyze the graph of the function $y = f(x) = x^3$ by finding: a) the x and y intercepts, b) the relative extrema, c) the points of inflection, d) the behavior near points where the function is not defined by finding the right and left hand limits at these points, and e) the behavior at infinity.

Solution: a) the curve passes through the origin, since (0, 0) satisfies the equation. There are no other x or y intercepts. b) The first derivative is equal to $3x^2$ and the second derivative is equal to $6x$. Since the first derivative is equal to 0 for $x = 0$, this is a possible extreme point. But since $3x^2$ is positive to the right and left of $x = 0$, (0, 0) is neither a maximum nor a minimum. c) The second derivative is negative to the left of 0 and positive to the right of 0 and equals 0 at $x = 0$. Therefore (0, 0) is a point of inflection. The second derivative also tells us that the graph is concave downward for $x < 0$ and concave upward for $x > 0$. d) The function is defined at all points. e) As $x \to -\infty$, $y \to -\infty$ and as $x \to +\infty$, $y \to +\infty$. The following plot confirms these properties.

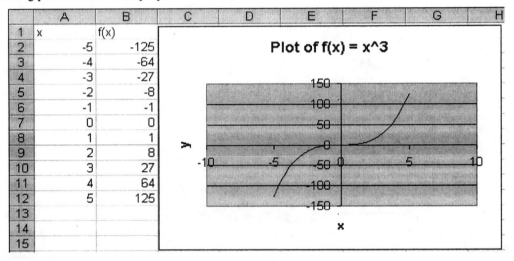

	A	B	C	D	E	F	G	H
1	x	f(x)						
2	-5	-125						
3	-4	-64						
4	-3	-27						
5	-2	-8						
6	-1	-1						
7	0	0						
8	1	1						
9	2	8						
10	3	27						
11	4	64						
12	5	125						
13								
14								
15								

10. Analyze the graph of the function $y = f(x) = x^4$ by finding: a) the x and y intercepts, b) the relative extrema, c) the points of inflection, d) the behavior near points where the function is not defined by finding the right and left hand limits at these points, and e) the behavior at infinity.

Solution: a) the curve passes through the origin, since (0, 0) satisfies the equation. There are no other x or y intercepts. b) The first derivative is equal to $4x^3$ and the second derivative is equal to $12x^2$. Since the first derivative is equal to 0 for $x = 0$, this is a possible extreme point. Since $4x^3$ is positive to the right of $x = 0$ and negative to the left of $x = 0$, (0, 0) is corresponds to a minimum on the graph. c) The second derivative is positive to the left of 0 and positive to the right of 0 and equals 0 at $x = 0$. Therefore (0, 0) is not a point of inflection. The second derivative also tells us that the graph is concave upward for $x < 0$ and concave upward for $x > 0$. d) The function is defined at all points. e) As $x \to -\infty$, $y \to +\infty$ and as $x \to +\infty$, $y \to +\infty$. The following plot confirms these properties:

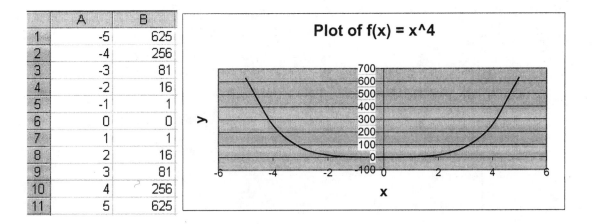

	A	B
1	-5	625
2	-4	256
3	-3	81
4	-2	16
5	-1	1
6	0	0
7	1	1
8	2	16
9	3	81
10	4	256
11	5	625

11. Analyze the graph of the function $y = f(x) = -3x^5 + 5x^3$ by finding: a) the x and y intercepts, b) the relative extrema, c) the points of inflection, d) the behavior near points where the function is not defined by finding the right and left hand limits at these points, and e) the behavior at infinity.

Solution: For part a), the curve passes through the origin, since (0, 0) satisfies the equation. There are also x-intercepts at $x = \pm \dfrac{\sqrt{15}}{3}$ or $x \approx \pm 1.29$. For part b), the first derivative is equal to $-15x^4 + 15x^2$ and the second derivative is equal to $-60x^3 + 30x$. The second derivative can be expressed as $-30x(2x^2 - 1)$. The first derivative can be expressed as $15x^2(1 - x^2)$. We see that the first derivative is equal to 0 when $x = -1, 0,$ or 1. The following is a plot of the first derivative.

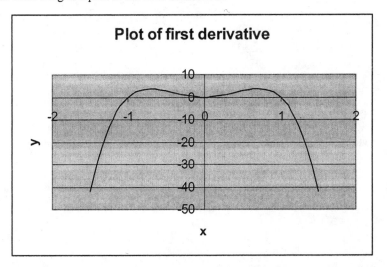

The first derivative is seen to change from negative to positive at $x = -1$. This indicates a relative minimum at the point $(-1, -2)$. The first derivative is positive on both sides of $x = 0$. This is not a relative minimum or relative maximum. The first derivative changes from positive to negative at $x = 1$. This indicates a relative maximum at the point $(1, 2)$. The following is a plot of the second derivative:

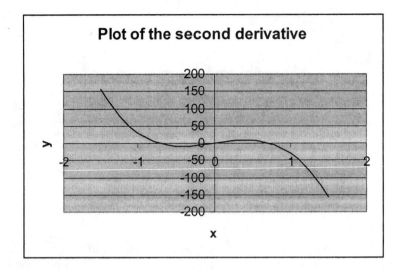

Plot of the second derivative

For part c), the second derivative is equal to $-30x(2x^2 - 1)$. The second derivative is equal to 0 when $x = 0$ and when $2x^2 - 1 = 0$ or $x = -\dfrac{\sqrt{2}}{2} = -0.707$ and when $x = 0.707$. There are three points of inflection located at $(-0.707, -1.237)$, $(0, 0)$, and $(0.707, 1.237)$. For part d), there are no points where the function is undefined. For part e), the following panel shows the behavior of the function at minus and plus infinity. As $x \to -\infty$, $y \to +\infty$ and as $x \to +\infty$, $y \to -\infty$.

	A	B	C	D	E
x	f(x)			x	f(x)
-10	295000			10	-295000
-100	3E+10			100	-3E+10
-1000	3E+15			1000	-3E+15
-10000	3E+20			10000	-3E+20

Note also from the graph of the second derivative that the second derivative is positive when $x < -0.707$, negative when $-0.707 < x < 0$, positive when $0 < x < 0.707$, and negative when $x > 0.707$. This tells us that the curve is concave upward when $x < -0.707$, concave downward when $-0.707 < x < 0$, concave upward when $0 < x < 0.707$, and concave downward when $x > 0.707$.

The following Excel plot of the function, $f(x) = -3x^5 + 5x^3$, illustrates the above discussion.

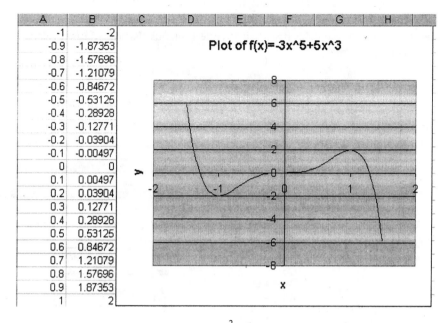

A	B
-1	-2
-0.9	-1.87353
-0.8	-1.57696
-0.7	-1.21079
-0.6	-0.84672
-0.5	-0.53125
-0.4	-0.28928
-0.3	-0.12771
-0.2	-0.03904
-0.1	-0.00497
0	0
0.1	0.00497
0.2	0.03904
0.3	0.12771
0.4	0.28928
0.5	0.53125
0.6	0.84672
0.7	1.21079
0.8	1.57696
0.9	1.87353
1	2

Plot of f(x)=-3x^5+5x^3

12. Analyze the graph of the function $y = f(x) = \dfrac{x^2 + 1}{x^2 - 1}$ by finding: a) the x and y intercepts, b) the relative extrema, c) the points of inflection, d) the behavior near points where the function is not defined by finding the right and left hand limits at these points, and e) the behavior at infinity.

Solution: For part a), there are no x intercepts since there are no values of x for which $y = 0$. The y intercept is at $(0, -1)$ since $y = -1$ when $x = 0$. For part b), the first derivative is $\dfrac{-4x}{(x^2 - 1)^2}$ and the second derivative is $\dfrac{4(3x^2 + 1)}{(x^2 - 1)^3}$. We see that the first derivative is equal to 0 when $x = 0$. Now plot the first derivative:

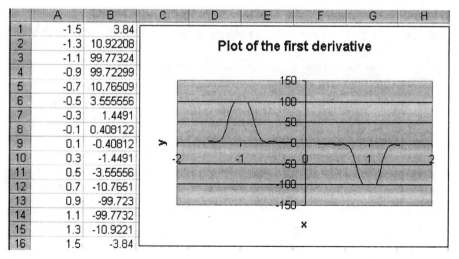

	A	B
1	-1.5	3.84
2	-1.3	10.92208
3	-1.1	99.77324
4	-0.9	99.72299
5	-0.7	10.76509
6	-0.5	3.555556
7	-0.3	1.4491
8	-0.1	0.408122
9	0.1	-0.40812
10	0.3	-1.4491
11	0.5	-3.55556
12	0.7	-10.7651
13	0.9	-99.723
14	1.1	-99.7732
15	1.3	-10.9221
16	1.5	-3.84

Plot of the first derivative

From the points on the left side of the graph, we see that the first derivative is positive to the left of $x = 0$ and negative to the right of $x = 0$. The first derivative test tells us that $(0, -1)$ is a relative maximum. Now plot of the second derivative:

	A	B	C	D	E	F	G	H
1	-1.5	15.872						
2	-1.3	73.9097						
3	-1.1	1999.784						
4	-0.9	-2000.29						
5	-0.7	-74.4812						
6	-0.5	-16.5926						
7	-0.3	-6.74124						
8	-0.1	-4.24611						
9	0.1	-4.24611						
10	0.3	-6.74124						
11	0.5	-16.5926						
12	0.7	-74.4812						
13	0.9	-2000.29						
14	1.1	1999.784						
15	1.3	73.9097						
16	1.5	15.872						

Plot of the second derivative

We see that $x = -1$ and $x = 1$ are asymptotes for this function. Let's look closer at its behavior between -0.5 and 0.5. This is shown in the following panel:

	A	B	C	D	E
1	-0.5	-16.5926		0	-4
2	-0.4	-9.98812		0.1	-4.24611
3	-0.3	-6.74124		0.2	-5.06366
4	-0.2	-5.06366		0.3	-6.74124
5	-0.1	-4.24611		0.4	-9.98812
6	0	-4		0.5	-16.5926

For part c), the second derivative is never equal to 0. There are no points of inflection. For part d), the function is undefined at $x = -1$ and $x = 1$. The following panel shows the behavior of the function near these points:

	A	B	C	D	E
1	x	f(x)		x	f(x)
2	-0.9	-9.526315789		0.9	-9.52632
3	-0.99	-99.50251256		0.99	-99.5025
4	-0.999	-999.5002501		0.999	-999.5
5	-0.9999	-9999.500025		0.9999	-9999.5
6	-1.1	10.52380952		1.1	10.52381
7	-1.01	100.5024876		1.01	100.5025
8	-1.001	1000.50025		1.001	1000.5
9	-1.0001	10000.50003		1.0001	10000.5

We see that the limit as x approaches -1 from the left is $+\infty$ and as x approaches -1 from the right is $-\infty$. The limit as x approaches 1 from the left is $-\infty$ and as x approaches 1 from the right is $+\infty$. For part e), the following panel shows the behavior of $f(x)$ at $\pm\infty$.

	A	B	C	D	E
1	x	f(x)		x	f(x)
2	-10	1.02020202		10	1.020202
3	-100	1.00020002		100	1.0002
4	-1000	1.000002		1000	1.000002
5	-10000	1.00000002		10000	1

We see that $f(x)$ approaches 1 as x goes to $+\infty$ as well as $-\infty$. The following Excel plot of $f(x) = \dfrac{x^2+1}{x^2-1}$

illustrates the above discussion. The graph is concave upward for $x < -1$, concave downward for $-1 < x < 1$, and concave upward for $x > 1$. This corresponds to the behavior of the graph of the second derivative. That is, the second derivative is positive for $x < -1$, negative for $-1 < x < 1$, and positive for $x > 1$.

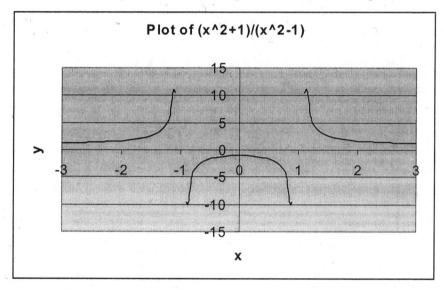

5-4 Related Rates

13. A cylindrical water tank is 50 feet high and has a radius equal to 15 feet. The tank supplies water to a small city. Water enters the tank at a constant rate, but the rate at which the residents use the water varies. When the tank is full, it contains $V = \pi r^2 h = \pi(225)(50) = 35{,}343$ cubic feet of water. When the water in the tank is at a height of h feet, the volume of water is equal to $V = 225\pi h$. The rate of change of the volume is related to the rate of change of height by the equation $\dfrac{dV}{dt} = 225\pi\dfrac{dh}{dt}$.

Make a table of the rate of changes of the volume for rate of changes of height equal to -0.1 to -2 in increments of 0.1 feet.

Solution: Excel has a built-in PI( ) function that contains no arguments and returns the value of π, accurate to 15 digits. Therefore, enter the values from –0.1 to –1 into column A and the formula, =225*PI()*A2, into cell B2. The values in column B are obtained by a click-and-drag. Use a similar procedure to obtain the values in columns C and D:

	A	B	C	D
1	rate of change of height	rate of change of volume	rate of change of height	rate of change of volume
2	-0.1	-70.686	-1.1	-777.544
3	-0.2	-141.372	-1.2	-848.230
4	-0.3	-212.058	-1.3	-918.916
5	-0.4	-282.743	-1.4	-989.602
6	-0.5	-353.429	-1.5	-1060.288
7	-0.6	-424.115	-1.6	-1130.973
8	-0.7	-494.801	-1.7	-1201.659
9	-0.8	-565.487	-1.8	-1272.345
10	-0.9	-636.173	-1.9	-1343.031
11	-1	-706.858	-2	-1413.717

The spreadsheet shows in row 11, for example, that a rate of change of height in the tank of –2 feet per hour causes a rate of change in volume of –1413.7 cubic feet per hour.

14. The volume of a spherical balloon is related to the radius of a balloon by the formula $V = \dfrac{4}{3}\pi r^3$.

The rate of change of the volume is related to the rate of change of the radius by the formula $\dfrac{dV}{dt} = 4\pi r^2 \dfrac{dr}{dt}$. Make a table that gives the rate of change of the radius for various times when the balloon is being filled at the rate of 4.5 cubic inches per minute.

Solution: The rate of change of the radius of the balloon is related to the rate of change of its volume by $\dfrac{dr}{dt} = \dfrac{1}{4\pi r^2}\dfrac{dV}{dt}$. Enter the values of time, 1 through 10, into column A. Since the balloon is being filled at 4.5 cubic inches per minute, the volume at time t is $4.5t$. Now enter the formula, =4.5*A2, into cell B2. The instantaneous radius of the balloon is related to the volume by the equation $r = \sqrt[3]{\dfrac{3V}{4\pi}}$. Enter the formula, =(3*B2/(4*PI()))^(1/3), into cell C2. Finally, since $\dfrac{dr}{dt} = \dfrac{1}{4\pi r^2}\dfrac{dV}{dt}$, enter the formula, =4.5/(4*PI()*C2^2), into cell D2: Click-and-drags return the required values:

	A	B	C	D
1	time	volume	radius	radius rate of change
2	1	4.5	1.024176	0.341392032
3	2	9	1.290381	0.215063503
4	3	13.5	1.477118	0.164124170
5	4	18	1.625778	0.135481518
6	5	22.5	1.751316	0.116754432
7	6	27	1.861051	0.103391748
8	7	31.5	1.959178	0.093294209
9	8	36	2.048352	0.085348008
10	9	40.5	2.130372	0.078902671
11	10	45	2.206521	0.073550684

We see that even though the volume is changing at a constant rate, the radius is increasing at a variable rate. The radius grows more and more slowly as time goes by.

5-5 Elasticity of Demand

15. Suppose the demand equation is $q = 100 - p^2$. Plot the demand function, the revenue function, and the elasticity function and discuss the results. Give that the revenue is at a maximum when the elasticity of demand is equal to 1, verify that this occurs when $p = \$5.77$.

Solution: The revenue function is $R = pq = p(100 - p^2) = 100p - p^3$. The elasticity of demand is defined as $E = -\dfrac{dq}{dp} \cdot \dfrac{p}{q} = -(-2p)\dfrac{p}{100 - p^2} = \dfrac{2p^2}{100 - p^2}$. First, let us consider the relationship between demand, revenue, elasticity, and price. To do this, enter prices from 0 to 9 in column A. Then in cell B2, enter the formula, =100-A2^2, in cell C2, enter the formula, =100*A2-A2^3, and in cell D2, enter the formula, =2*A2^2/(100-A2^2). Click-and-drags produce the results in the following panel.

	A	B	C	D
1	p	q	R	E
2	0	100	0	0
3	1	99	99	0.020202
4	2	96	192	0.083333
5	3	91	273	0.197802
6	4	84	336	0.380952
7	5	75	375	0.666667
8	6	64	384	1.125
9	7	51	357	1.921569
10	8	36	288	3.555556
11	9	19	171	8.526316

Note that as the price increases, the daily demand goes down. As the price increases, the revenue increases but between $5 and $6 a maximum revenue is reached. As the price increases thereon, the revenue begins to decrease. It is also observed that the elasticity of demand equals 1 somewhere between $5 and $6. We know from the discussion above that this occurs when the price equals $5.77. The demand is inelastic when the price is less than $5.77, has unit elasticity when the price equals $5.77, and is elastic when the price is greater than $5.77. Use the Chart Wizard to obtain the plots of demand versus price, revenue versus price, and elasticity versus price:

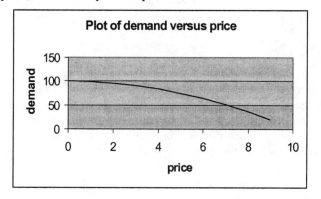

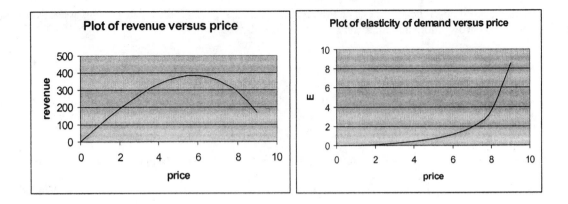

16. Suppose the demand equation is $q = 100e^{-3p^2+p}$. Plot the demand function, the revenue function, and the elasticity function. Interpret your results. Given that the revenue is at a maximum when the elasticity of demand is equal to 1, verify that this occurs when $p = 0.5$.

Solution: The revenue function is $R = pq = p100e^{-3p^2+p} = 100pe^{-3p^2+p}$. The elasticity of demand is given by $E = -\dfrac{dq}{dp} \cdot \dfrac{p}{q} = 6p^2 - p$. First, let us consider the relationship between demand, revenue, elasticity, and price. To do this, enter the prices from 0 to 1 in column A. In cell B2, enter the formula, =100*EXP(-3*A2^2+A2), in cell C2, enter the formula, =100*A2*(EXP(-3*A2^2+A2)), and in cell D2, enter the formula, =6*A2^2-A2. Click-and-drags produce the results in the following panel:

	A	B	C	D
1	p	q	R	E
2	0	100	0	0
3	0.1	107.2508	10.72508	-0.04
4	0.2	108.3287	21.66574	0.04
5	0.3	103.0455	30.91364	0.24
6	0.4	92.31163	36.92465	0.56
7	0.5	77.88008	38.94004	1
8	0.6	61.87834	37.127	1.56
9	0.7	46.30131	32.41091	2.24
10	0.8	32.62798	26.10238	3.04
11	0.9	21.65357	19.48821	3.96
12	1	13.53353	13.53353	5

Note that as the price increases, the daily demand first increases and then goes back down. As the price increases, the revenue increases until $p = 0.5$. The maximum revenue is reached and then as the price increases from $p = 0.5$, the revenue begins to decrease. It is also observed that the elasticity of demand equals 1 when $p = 0.5$. The demand is inelastic when the price is less than 0.5, has unit elasticity when the price equals 0.5, and is elastic when the price is greater than 0.5. The plots of demand versus price, revenue versus price, and elasticity versus price are obtained using the chart master in the usual manner and are as follows:

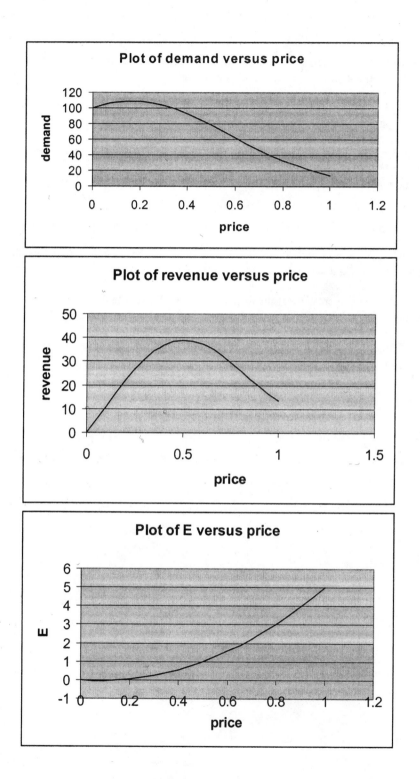

Chapter Review Exercises

1. Find the location of all the relative and absolute extrema of the function $f(x) = x^2 - 4x + 1$ with the domain $[0, 3]$. Use graphical as well as calculus techniques.

2. Find the location of all the relative and absolute extrema of the function $f(x) = x^3 + x$ with the domain $[-2, 2]$. Use graphical as well as calculus techniques.

3. Bill Brown wants to fence in a rectangular vegetable patch. The fencing for the east and west sides costs $4, while the fencing for the north and south sides costs only $2 per foot. If Bill's budget is $80 for the project, what is the largest area that he can enclose?

4. Analyze the graph of the function $y = f(x) = \dfrac{x^2 - 1}{x^2 + 1}$ by finding: a) the x and y intercepts, b) the relative extrema, c) the points of inflection, d) the behavior near points where the function is not defined by finding the right and left hand limits at these points, and e) the behavior at infinity.

5. The radius of a circle is increasing at the rate of 5 cm/sec. Make a table showing how fast the area is increasing at the instant when the radius of the circle has reached the values $1, 1.5, \ldots, 10$.

6. Suppose the demand equation is $q = 1000 - 10p$. Plot the demand function, the revenue function, and the elasticity function and discuss the results.

CHAPTER 6

The Integral

6-1 The Indefinite Integral

Antiderivatives

1. The power rule, the sum and difference rules, and the constant multiple rule applied to the function

$f(x) = 2x^3 - 4x + 3$ give $F(x) = \dfrac{2x^4}{4} - \dfrac{4x^2}{2} + 3x + C$ for the antiderivative of $f(x)$. We express this

result as follows: $F(x) = \displaystyle\int 2x^3 - 4x + 3 \ dx = \dfrac{2x^4}{4} - \dfrac{4x^2}{2} + 3x + C$. We also have the following

result for every value of x and any value of C: $\dfrac{dF(x)}{dx} = f(x)$. To get a feel for these results:

a) Let $C = 10$ and then evaluate $\dfrac{dF(x)}{dx}$ at $x = 5$ using the limit definition of the derivative and show

  that the result is equal to $f(5)$.

b) Let $C = 100$ and then evaluate $\dfrac{dF(x)}{dx}$ at $x = 7$ using the limit definition of the derivative and show

  that the result is equal to $f(7)$.

Solution: For part a) of the problem, use the limit definition, $\displaystyle\lim_{h \to 0} \dfrac{F(5+h) - F(5-h)}{2h}$, and the technique

illustrated in section 3-3. We arrive at the following numerical evaluation of the derivative of $F(x)$ when $x = 5$. Build a spreadsheet by first entering the usual values of h are into column A. Then enter the formula, =0.5*(5+A2)^4-2*(5+A2)^2+3*(5+A2)+10, into cell B2, the formula, =0.5*(5-A2)^4-2*(5-A2)^2+3*(5-A2)+10, into cell C2, and the formula, =(B2-C2)/(2*A2), into cell D2. Click-and-drags produce the following spreadsheet:

	A	B	C	D
1	h	F(5+h)	F(5-h)	(F(5+h)-F(5-h))/2h
2	1	604	118	243
3	0.1	311.5401	264.9201	233.1
4	0.01	289.8373	285.1773	233.001
5	0.001	287.7331	287.2671	233.00001
6	0.0001	287.5233	287.4767	233.0000001

Therefore, the limit is equal to 233. The same result is obtained by evaluating $f(5) = 2(125) - 4(5) + 3 =$

233. We have illustrated that if $F(x) = \dfrac{2x^4}{4} - \dfrac{4x^2}{2} + 3x + 10$, then $\dfrac{dF(x)}{dx} = f(x)$ when $x = 5$. It is

recognized that this method is not precisely a proof that the antiderivative is correct, but nevertheless it is an illustration of the result. We shall use a similar procedure to solve subsequent problems.

For part b) of the problem, use the limit definition, $\lim\limits_{h \to 0} \dfrac{F(7+h) - F(7-h)}{2h}$, and the same technique as in part a). We arrive at the following numerical evaluation of the derivative of $F(x)$ when $x = 7$. The correct formulas are $=0.5*(7+A2)^4-2*(7+A2)^2+3*(7+A2)+100$, for cell B2, $=0.5*(7-A2)^4-2*(7-A2)^2+3*(7-A2)+100$ for cell C2, and $=(B2-C2)/(2*A2)$ for cell D2. Click-and-drags produce the following spreadsheet:

	A	B	C	D
	h	F(7+h)	F(7-h)	(F(7+h)-F(7-h))/2h
1				
2	1	2044	694	675
3	0.1	1291.064	1158.836	661.14
4	0.01	1230.125	1216.904	661.0014
5	0.001	1224.161	1222.839	661.000014
6	0.0001	1223.566	1223.434	661.0000001

The limit is equal to 661. The same result is obtained by evaluating $f(7) = 2(343) - 4(7) + 3 = 661$. This procedure illustrates that if $F(x) = \dfrac{2x^4}{4} - \dfrac{4x^2}{2} + 3x + 100$, then $\dfrac{dF(x)}{dx} = f(x)$ when $x = 7$.

Acceleration, Velocity and Position

2. The velocity of a particle moving in a straight line in a linear accelerator is given by $v(t) = 3e^t + t$. Find an expression for the position after time t. Given that $s = 0$ when $t = 0$, find the constant of integration. Plot the velocity and distance function for $0 < t < 3$.

Solution: The distance function $s(t)$ is the antiderivative of $v(t)$. $s(t) = 3e^t + t^2/2 + C$. To evaluate the constant of integration, we set $s(0) = 3e^0 + C = 0$ and obtain $C = -3$. Therefore, $s(t) = 3e^t + t^2/2 - 3$. Enter the formula, $=3*EXP(A2)+A2$, into cell B2 and the formula, $=3*EXP(A2)+(A2^2)/2-3$, into cell C2. Excel returns the values required for plotting after performing a click-and-drag. The graphs are plotted using the Chart Wizard in the same manner as illustrated in previous chapters. The velocity is shown as a solid curve and the distance is shown as a dashed line:

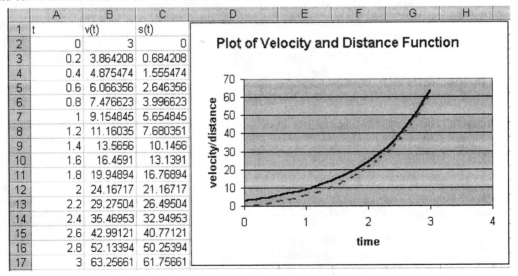

	A	B	C
1	t	v(t)	s(t)
2	0	3	0
3	0.2	3.864208	0.684208
4	0.4	4.875474	1.555474
5	0.6	6.066356	2.646356
6	0.8	7.476623	3.996623
7	1	9.154845	5.654845
8	1.2	11.16035	7.680351
9	1.4	13.5656	10.1456
10	1.6	16.4591	13.1391
11	1.8	19.94894	16.76894
12	2	24.16717	21.16717
13	2.2	29.27504	26.49504
14	2.4	35.46953	32.94953
15	2.6	42.99121	40.77121
16	2.8	52.13394	50.25394
17	3	63.25661	61.75661

Plot of Velocity and Distance Function

6-2 Substitution

3. The method of substitution can be used to find the indefinite integral $\int \sqrt{1-3x}\,dx$. Choose $u = 1 - 3x$.

The differential du is given by $du = -3dx$. This implies that dx is given by $dx = -\frac{1}{3}du$. Substituting

we obtain $\int \sqrt{1-3x}\,dx = \int -\frac{1}{3}u^{1/2}du = -\frac{2}{9}u^{3/2} + C$. The antiderivative in terms of x is defined as

$F(x) = \int \sqrt{1-3x}\,dx = -\frac{2}{9}(1-3x)^{3/2} + C$. We also have the following result for $x < 1/3$ and for any

value of C: $\frac{dF(x)}{dx} = f(x)$, where $f(x) = \sqrt{1-3x}$. Let $C = 0$ and then evaluate $\frac{dF(x)}{dx}$ at $x = -5$

using the limit definition of the derivative. Then show that the result is equal to $f(-5)$.

Solution: Using the limit definition, $\lim\limits_{h \to 0} \dfrac{F(-5+h) - F(-5-h)}{2h}$, and the technique illustrated in section

3-3, we arrive at the following numerical evaluation of the derivative of $F(x)$ when $x = -5$. Enter the formula, =(-2/9)*(1-3*(-5+A2))^1.5, into cell B2, the formula, =(-2/9)*(1-3*(-5-A2))^1.5, into cell C2, and the formula, =(B2-C2)/(2*A2), into cell D2. Click-and-drags produce the following spreadsheet:

	A	B	C	D
1	h	F(-5+h)	F(-5-h)	(F(-5+h)-F(-5-h))/2h
2	1	-10.416	-18.4042	3.994101483
3	0.1	-13.8241	-14.6241	3.999941402
4	0.01	-14.1822	-14.2622	3.999999414
5	0.001	-14.2182	-14.2262	3.999999994
6	0.0001	-14.2218	-14.2226	4

The same result is obtained by evaluating $f(-5) = \sqrt{1-3(-5)} = 4$. This procedure illustrates that if

$F(x) = -\frac{2}{9}(1-3x)^{3/2}$, then $\frac{dF(x)}{dx} = f(x)$ when $x = -5$.

6-3 The Definite Integral as a Sum: A Numerical Approach

4. Calculate the Riemann sum for the integral $\int\limits_{0}^{1} \dfrac{1}{1+x}\,dx$ using $n = 5$ subdivisions.

Solution: The integrand is the function $f(x) = \dfrac{1}{1+x}$, the limits of integration are 0 and 1, and x is the

variable of integration. Since the interval is [0, 1] and $n = 5$, $\Delta x = 1/5 = 0.2$. The subdivisions of [0, 1] are $0 < 0.2 < 0.4 < 0.6 < 0.8 < 1$. The Riemann sum is $[f(0) + f(0.2) + f(0.4) + f(0.6) + f(0.8)](0.2)$. Enter the values 0, 0.2, 0.4, 0.6, and 0.8 into column A. Then enter the expression, =1/(1+A1), into cell B1 and fill out the required values in column B by performing a click-and-drag. Now enter the formula, =SUM(B1:B5)*0.2, into cell C1. The following panel gives the results:

	A	B	C
1	0	1	0.745635
2	0.2	0.833333	
3	0.4	0.714286	
4	0.6	0.625	
5	0.8	0.555556	

The Riemann sum is shown in cell C1 to be 0.745635.

5. Calculate the Riemann sum for the integral $\int_0^2 (-x^2 + 5)dx$ using a) $n = 10$, b) $n = 100$, c) $n = 1000$.

Solution: For part a), since the interval is $[0, 2]$ and $n = 10$, $\Delta x = 2/10 = 0.2$. The subdivisions of $[0, 2]$ are $0 < 0.2 < 0.4 < \cdots < 2$. The Riemann sum we want is $[f(0) + f(0.2) + \cdots + f(1.8)](0.2)$. Enter the subdivision points into column A and the formula, =-(A1^2)+5, into cell B1. Perform a click-and-drag to fill out the required values in column B. Now enter the formula, =SUM(B1:B10)*0.2, into cell C1. The following panel gives the results:

	A	B	C
1	0	5	7.72
2	0.2	4.96	
3	0.4	4.84	
4	0.6	4.64	
5	0.8	4.36	
6	1	4	
7	1.2	3.56	
8	1.4	3.04	
9	1.6	2.44	
10	1.8	1.76	

The Riemann sum is shown in cell C1 to be 7.72. (Note that you might think it reasonable to use the formula, =-A1^2+5, rather than =-(A1^2)+5. As explained in chapter 1, Excel processes the negation prior to the exponentiation. Therefore, parentheses must be supplied.)

For part b), since the interval is $[0, 2]$ and $n = 100$, $\Delta x = 2/100 = 0.02$. The subdivisions of $[0, 2]$ are $0 < 0.02 < 0.04 < \cdots < 2$. The Riemann sum we want is $[f(0) + f(0.02) + \cdots + f(1.98)](0.02)$. Rather than the usual click-and-drag, Excel offers a powerful tool for entering subdivision points whenever the number of subdivisions is large. To enter the subdivision points into column A, first enter the first point (0 in this case) into cell A1. Select this cell and from the Edit menu, select Fill → Series… In the dialog box, choose the Series in Columns option, the Type Linear option, and enter 0.02 as the Step value (Δx), and 1.98 as the Stop value:

Excel will fill in the cells A1:A100 with the correct values. Now enter the formula, =-(A1^2)+5, into cell B1 and a click-and-drag to fill out the required values in column B. Then enter the formula, =SUM(B1:B100)*0.02, into cell C1. The following panel gives the results.

	A	B	C			A	B
1	0	5	7.3732		91	1.8	1.76
2	0.02	4.9996			92	1.82	1.6876
3	0.04	4.9984			93	1.84	1.6144
4	0.06	4.9964			94	1.86	1.5404
5	0.08	4.9936			95	1.88	1.4656
6	0.1	4.99			96	1.9	1.39
7	0.12	4.9856			97	1.92	1.3136
8	0.14	4.9804			98	1.94	1.2364
9	0.16	4.9744			99	1.96	1.1584
10	0.18	4.9676		...	100	1.98	1.0796

The Riemann sum is shown in cell C1 to be 7.3732.

For part c), since the interval is [0, 2] and $n = 1000$, $\Delta x = 2/1000 = 0.002$. The subdivisions of [0, 2] are $0 < 0.002 < 0.004 < \cdots < 2$. The Riemann sum we want is $[f(0) + f(0.002) + \cdots + f(1.998)](0.002)$. To enter the subdivision points into column A use a procedure similar to that used in part b). Enter the first point, 0 in this case, into cell A1. Select this cell and from the Edit menu, select Fill $\rightarrow$ Series... In the dialog box, choose Series in Columns, Type Linear, and enter 0.002 as the Step value (Δx), and 1.998 as the Stop value. Excel will fill in the cells A1:A1000 with all of the required values. Now enter the formula, =-(A1^2)+5, into cell B1 and perform a click-and-drag to fill out the required values in column B. Enter the formula, =SUM(B1:B1000)*0.002, into cell C1. The following panel gives the results:

	A	B	C			A	B
1	0	5	7.337332		991	1.98	8.9204
2	0.002	4.999996			992	1.982	8.928324
3	0.004	4.999984			993	1.984	8.936256
4	0.006	4.999964			994	1.986	8.944196
5	0.008	4.999936			995	1.988	8.952144
6	0.01	4.9999			996	1.99	8.9601
7	0.012	4.999856			997	1.992	8.968064
8	0.014	4.999804			998	1.994	8.976036
9	0.016	4.999744			999	1.996	8.984016
10	0.018	4.999676		...	1000	1.998	8.992004

The Riemann sum is shown in cell C1 to be 7.337332. Observe that the Riemann sums are approaching a limiting value. We shall see later that this limiting value is actually equivalent to the area under the graph of $f(x) = -x^2 + 5$ for x between $x = 0$ and $x = 2$ and is equal to 22/3.

6. The distribution of adult female heights is modeled by the function, $f(x) = \dfrac{1}{\sqrt{2\pi}(2.5)} e^{-\frac{(x-65)^2}{2(2.5)^2}}$. The

percent of females with heights between 60 and 70 inches is given by $\int\limits_{60}^{70} f(x)dx$. Use a Riemann sum

with $n = 50$ to estimate the percent of females with heights between 60 and 70 inches. Then plot the data used to obtain the sum.

Solution: Fill in the series of values from 60 to 70 at intervals of $\Delta x = 0.2$ as described in the preceding problem. Then enter the formula, =1/(SQRT(2*PI())*2.5)*EXP(-((A1-65)^2/12.5)), into cell B1 and Excel will return the required function values with a click-and-drag. Finally, enter the formula, =SUM(B1:B50)*0.2, into cell C1, and Excel returns the Riemann sum, 0.954385:

	A	B	C
1	60	0.021596	0.954385
2	60.2	0.025263	
3	60.4	0.029363	
4	60.6	0.033911	
5	60.8	0.038913	

	A	B
46	69	0.044368
47	69.2	0.038913
48	69.4	0.033911
49	69.6	0.029363
50	69.8	0.025263

We see that 95.4% of the females are between 60 and 70 inches tall. The plot using the Chart Wizard is shown in the following graph:

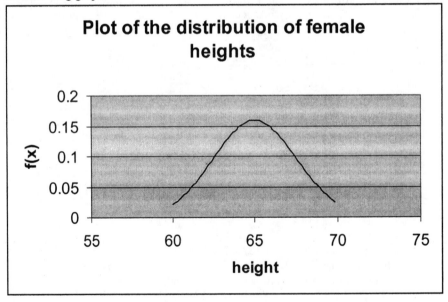

The curve is bell-shaped. We can conclude that female heights are normally distributed with a mean of 65 inches. We also know that over 95% have heights between 60 and 70 inches.

6-4 The Definite Integral as Area: A Geometric Approach

7. Use formulas from geometry to find the areas of: a) a rectangle of length 4 and height 2 and b) a right triangle having base equal to 4 and height equal to 4. Use Riemann sums with $n = 10$ to approximate these areas.

Solution: The following graphs show the areas in question:

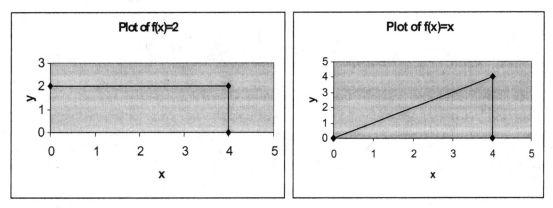

a) The area enclosed by the x-axis, the y-axis, the line $f(x) = 2$, and the $x = 4$ is a 2 by 4 rectangle. The area is $A = l(w) = 4(2) = 8$ square units. Using a Riemann sum with $n = 10$, we have $\Delta x = 4/10 = 0.4$. The subdivision points are 0, 0.4, $\cdots$, 4. The Riemann sum is as follows:

$$[f(0) + f(.4) + \cdots f(3.6)](0.4) = [2 + 2 + , \cdots, + 2](0.4) = (20)(0.4) = 8$$

In this case, the Riemann sum and the formula for the area give the same result. So in this case the Riemann sum gives the exact answer for the area.

b) The area enclosed by the x-axis, the y-axis, the line $f(x) = x$, and the $x = 4$ is a triangle with base $b = 4$ and height $h = 4$. The area is $A = .5bh = .5(4)(4) = 8$ square units. Using a Riemann sum approximation with $n = 10$, we have $\Delta x = 4/10 = 0.4$. The subdivision points are 0, 0.4, $\cdots$, 4. The left-hand Riemann sum is

$$[f(0) + f(.4) + \cdots f(3.6)](0.4) = [0 + .4 + , \cdots, + 3.6](0.4)$$

The right-hand Riemann sum is

$$[f(.4) + \cdots f(4)](0.4) = [0.4 + , \cdots, +4](0.4)$$

The mid-point Riemann sum is

$$[f(.2) + \cdots f(3.8)](0.4) = [0.2 + , \cdots, + 3.8](0.4)$$

The following panel gives the left-hand, right-hand, and midpoint Riemann sums for the area.

	A	B	C	D	E	F
1	0	7.2	0.4	8.8	0.2	8
2	0.4		0.8		0.6	
3	0.8		1.2		1	
4	1.2		1.6		1.4	
5	1.6		2		1.8	
6	2		2.4		2.2	
7	2.4		2.8		2.6	
8	2.8		3.2		3	
9	3.2		3.6		3.4	
10	3.6		4		3.8	

Enter the formula, =SUM(A1:A10)*0.4, into cell B1, the formula, =SUM(C1:C10)*0.4, into cell D1, and the formula, =SUM(E1:E10)*0.4, into cell F1. The left-hand Riemann sum evaluates the function at the left-hand side of each interval, the right-hand Riemann sum evaluates the function at the right-hand side of each interval, and the mid-point Riemann sum evaluates the function at the midpoint of each interval. In this problem, the left-hand Riemann sum underestimates the area by 0.8 square units, the right-hand Riemann sum overestimates the area by 0.8 square units, and the mid-point Riemann sum equals the area.

8. Use Riemann sums with $n = 10$ and $n = 100$ to approximate the area represented by $\displaystyle\int_{0}^{\pi/4} \tan x \; dx$.

Construct a graph of the area you are finding.

Solution: First, construct the graph as follows. Note first that $\pi/4 \approx 0.79$. Full in the values from 0 to 0.8 at intervals of 0.05 in column A. Then enter the formula, =TAN(A1), into cell B1. The usual click-and-drag returns the required values in column B. Finally, use the Chart Wizard to construct the graph of the tangent function for x between $x = 0$ and $x = \pi/4$:

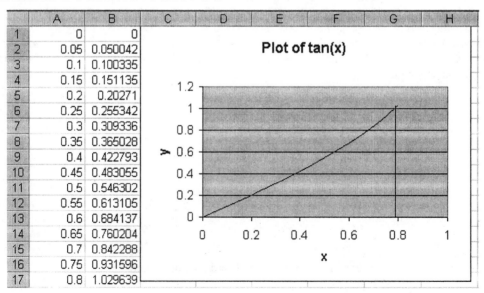

	A	B
1	0	0
2	0.05	0.050042
3	0.1	0.100335
4	0.15	0.151135
5	0.2	0.20271
6	0.25	0.255342
7	0.3	0.309336
8	0.35	0.365028
9	0.4	0.422793
10	0.45	0.483055
11	0.5	0.546302
12	0.55	0.613105
13	0.6	0.684137
14	0.65	0.760204
15	0.7	0.842288
16	0.75	0.931596
17	0.8	1.029639

Plot of tan(x)

The area enclosed by the tangent curve, the line $x = 0.79$ and the x-axis is represented by $\int_0^{\pi/4} \tan x \; dx$.

For $n = 10$, $\Delta x = 0.8/10 = 0.08$. The Riemann sum approximation to the area is

$$[f(0) + f(0.08) + \cdots + f(0.72)](0.08) = [\tan(0) + \tan(0.08) + \cdots + \tan(0.72)](0.08)$$

Fill in the left-hand endpoints of the intervals in column A. Now enter the formula, =TAN(A1), into cell B1 and the tangent values are returned in column B. Finally, enter the formula, =SUM(B1:B10)*0.08, into cell C1. Excel returns the required value, 0.32077.

	A	B	C
1	0	0	0.32077
2	0.08	0.080171	
3	0.16	0.161379	
4	0.24	0.244717	
5	0.32	0.331389	
6	0.4	0.422793	
7	0.48	0.520611	
8	0.56	0.62695	
9	0.64	0.744544	
10	0.72	0.877068	

A similar procedure is performed for $n = 100$.

	A	B	C
1	0	0	0.357278
2	0.008	0.008	
3	0.016	0.016001	
4	0.024	0.024005	
5	0.032	0.032011	
6	0.04	0.040021	
7	0.048	0.048037	
8	0.056	0.056059	
9	0.064	0.064088	
10	0.072	0.072125	

	A	B
91	0.72	0.877068
92	0.728	0.891322
93	0.736	0.905781
94	0.744	0.920451
95	0.752	0.935339
96	0.76	0.950451
97	0.768	0.965795
98	0.776	0.981378
99	0.784	0.997208
100	0.792	1.013292

For $n = 100$ subdivisions, the approximation to the area is 0.357278. The exact area is equal to 0.347. If you repeat the above procedure for $n = 1000$, you will get closer to this value. In section 6-5, we learn how to find the exact area.

9. Plot the function $f(x) = \sin x$ for $0 < x < 2\pi$. Then approximate $\int_0^{2\pi} \sin x \; dx$ by using a Riemann sum with $n = 10$ subdivisions. Explain your results.

Solution: The plot is obtained in the usual way. Fill in the values from 0 to 6.3 (approximately 2π) in column A. Then enter the formula, =SIN(A1), into cell B1 to obtain the required values for plotting. Use the Chart Wizard to produce the plot:

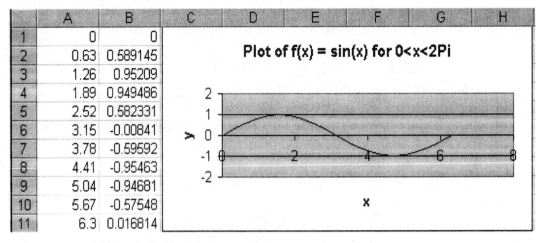

	A	B	C	D	E	F	G	H
1	0	0						
2	0.63	0.589145						
3	1.26	0.95209						
4	1.89	0.949486						
5	2.52	0.582331						
6	3.15	-0.00841						
7	3.78	-0.59592						
8	4.41	-0.95463						
9	5.04	-0.94681						
10	5.67	-0.57548						
11	6.3	0.016814						

For $n = 10$, $\Delta x = 6.3/10 = 0.63$. The Riemann sum is

$$[\sin(0) + \sin(0.63) + \sin(1.26) + \cdots + \sin(5.67)](0.63).$$

When the entries in column B are summed and multiplied by 0.63, the result is –0.00516. The reason that the value is close to 0 is that $\int_0^{2\pi} \sin x \, dx$ = area above the x-axis minus area below the x-axis = 0.

10. The rate of spending over an 8-year period by Medicare on hospice care in the U.S., in billions of dollars per year, is given by $f(x) = \dfrac{2.2e^{0.67x}}{13 + e^{0.67x}}$, where $0 < x < 8$. The total amount spent over the 8-year period is represented by $\int_0^8 \dfrac{2.2e^{0.67x}}{13 + e^{0.67x}} \, dx$. Graph the function, $f(x)$, and estimate the total amount spent by use of a Riemann sum with $n = 10$.

Solution: When the 8-year period is divided into 10 subdivisions, $\Delta x = 8/10 = 0.8$. The Riemann sum is

$$[f(0) + f(0.8) + \cdots + f(7.2)](0.8).$$

Use column A to fill in the subdivision points. The formula, =2.2*EXP(0.67*A1)/(13+EXP(0.67*A1)), entered into cell B1 returns the required values for obtaining the sum. The formula for the sum, =SUM(B1:B10)*0.8, entered into cell C1 returns the value 8.361804. The graph is also shown in the next panel. According to the model, Medicare spent approximately $8.36 billion on hospice care over the 8-year period.

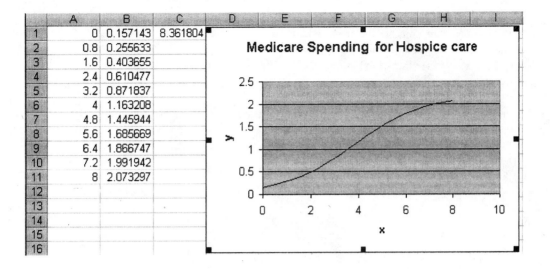

	A	B	C	D	E	F	G	H	I
1	0	0.157143	8.361804						
2	0.8	0.255633							
3	1.6	0.403655							
4	2.4	0.610477							
5	3.2	0.871837							
6	4	1.163208							
7	4.8	1.445944							
8	5.6	1.685669							
9	6.4	1.866747							
10	7.2	1.991942							
11	8	2.073297							
12									
13									
14									
15									
16									

6-5 The Definite Integral: An Algebraic Approach and the Fundamental Theorem of Calculus

Computing the Area Function Using Geometry

11. Use geometry to compute $A(x) = \int_a^x f(t)dt$ and verify that $A'(x) = f(x)$. Also find $A(x)$ using the fundamental theorem of calculus in each of the following instances:

a) $A(x) = \int_0^x 2t \ dt$

b) $A(x) = \int_3^x 5 \ dt$

c) $A(x) = \int_1^x 3t \ dt$

Solution:

a) The following Excel graph, created using the Chart Wizard, represents a segment of $f(t) = 2t$ for $0 < t < 2$:

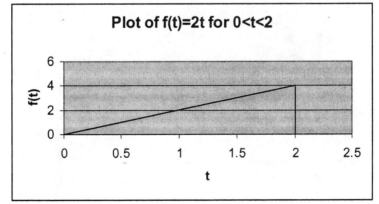

The area inside the triangle bounded by segments of the lines $f(t) = 2t$, $t = 2$, and the t-axis is $A(2) = 0.5(2)(4) = 4$ square units. If the line $t = 2$ is replaced by the line $t = x$, the base of the triangle becomes x units long and the altitude of the triangle becomes $2x$ units long. The new area is $A(x) = 0.5(x)(2x) = x^2$. We see that $A'(x) = 2x = f(x)$. By the fundamental theorem of calculus, $A(x) = F(x) - F(0)$, where $F(x) = x^2$, or $A(x) = x^2 - 0^2 = x^2$.

b) The following Excel graph, created using the Chart Wizard, represents a segment of the function, $f(t) = 5$ for $3 < t < 6$.

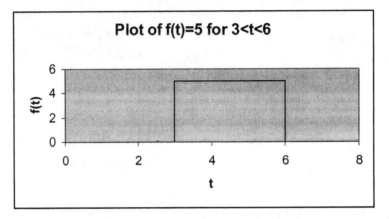

The area inside the rectangle bounded by segments of the lines $f(t) = 5$, $t = 3$, $t = 6$, and the t-axis is $A(6) = 3(5) = 15$ square units. If the line $t = 6$ is replaced by the line $t = x$, $x > 3$, the width of the rectangle becomes $x - 3$. The length remains at 5. The area is $A(x) = 5(x - 3) = 5x - 15$ and $A'(x) = 5 = f(x)$. By the fundamental theorem of calculus, $A(x) = F(x) - F(3)$, where $F(x) = 5x$. $A(x) = 5x - 5(3) = 5x - 15$.

c) The following Excel graph, created using the Chart Wizard, represents a segment of the function, $f(t) = 3t$ for $1 < t < 4$.

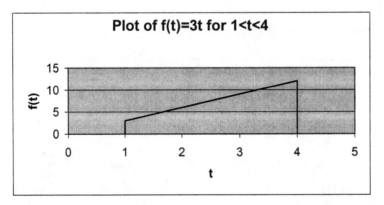

The trapezoid bounded by the segments of the lines $f(t) = 3t$, $t = 1$, $t = 4$, and the t-axis has area, $.5(3 + 12)(3) = 22.5$. If the line $t = 4$ is replaced by the line $t = x$, the area is $A(x) = .5(3 + 3x)(x - 1) = .5(3x^2 - 3) = 1.5x^2 - 1.5$ and $A'(x) = 3x$. By the fundamental theorem of calculus, $A(x) = F(x) - F(1)$, where $F(x) = 1.5x^2$. Evaluating: $A(x) = 1.5x^2 - 1.5(1)^2 = 1.5x^2 - 1.5$.

Comparing Riemann Approximations with Exact Values of Definite Integrals

12. Find the approximate values of $\int_0^x t^2 dt$ for x varying from 0 to 1 at intervals of 0.1 using Riemann sum approximations and compare these with the exact values obtained by using the fundamental theorem of calculus.

Solution: Fill in the numbers 0 through 1, at intervals of 0.1, in column A. Enter the formula, =A1^2, in cell B1 and execute a click-and-drag to obtain the required squares of the values in column A. Then enter the formula, =0.1*SUM(B$1:B1), into cell C1. The usual click-and-drag performed in column C returns the values for =0.1*SUM(B$1:B2) to cell C2, =0.1*SUM(B$1:B3) to cell C3, and so forth until the value for =0.1*SUM(B$1:B11) is returned to cell C11. The required Riemann sums are now in column C:

	A	B	C	D
1	0	0	0	0
2	0.1	0.01	0.001	0.000333
3	0.2	0.04	0.005	0.002667
4	0.3	0.09	0.014	0.009
5	0.4	0.16	0.03	0.021333
6	0.5	0.25	0.055	0.041667
7	0.6	0.36	0.091	0.072
8	0.7	0.49	0.14	0.114333
9	0.8	0.64	0.204	0.170667
10	0.9	0.81	0.285	0.243
11	1	1	0.385	0.333333

The exact values obtained by using the fundamental theorem of calculus are found by noting that $\int_0^x t^2 dt$

is equal to $F(x) - F(0)$, where $F(x) = \dfrac{x^3}{3}$. These are now found by entering the formula, =A1^3/3, into

cell D1 and performing a click-and-drag operation. Summarizing, the approximations to $\int_0^x t^2 dt$, where x

takes on the values in column A are in column C and the exact values for $\int_0^x t^2 dt$, where x takes on the

values in column A are in column D.

Computing Area

13. Use the fundamental theorem of calculus to compute $\int_0^2 e^{x^3}(3x^2)dx$ and graph the area that this

integral represents.

Solution: The antiderivative of the integrand is $F(x) = e^{x^3}$. The value of the integral is $F(2) - F(0) = e^8$ $- e^0 = 2980.96 - 1 = 2979.96$. The function $f(x) = 3x^2 e^{x^2}$ is a function that grows very rapidly. When x = 2, the value of the function is 35,771.5. Therefore the function grows from 0 when x = 0 to 35,771.5 when x = 2. To graph the function using Excel, enter the numbers from 0 to 2 by intervals of 0.1 in column A. Then enter the formula, =3*A1^2*EXP(A1^3), into cell B1 and the function values are returned to column B by using a click-and-drag operation. After these values are created, enter a final point (2, 0) by using cells A22 and B22. The purpose of this point is to tie the curve down on the right side. The area that is contained between the segments of the curve $f(x) = 3x^2 e^{x^3}$, $x = 2$, and the x-axis is

equal to $\int_0^2 e^{x^3}(3x^2)dx = 2979.96$. The following panel shows the graph for $0 < x < 2$ and a segment of

the line x = 2. It is clear from the graph that the majority of the area under the curve occurs to the right of x = 1.5.

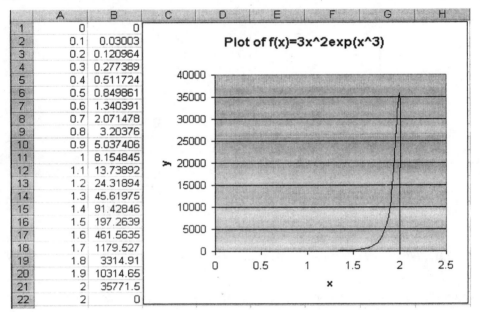

	A	B
1	0	0
2	0.1	0.03003
3	0.2	0.120964
4	0.3	0.277389
5	0.4	0.511724
6	0.5	0.849861
7	0.6	1.340391
8	0.7	2.071478
9	0.8	3.20376
10	0.9	5.037406
11	1	8.154845
12	1.1	13.73892
13	1.2	24.31894
14	1.3	45.61975
15	1.4	91.42846
16	1.5	197.2639
17	1.6	461.5635
18	1.7	1179.527
19	1.8	3314.91
20	1.9	10314.65
21	2	35771.5
22	2	0

Plot of f(x)=3x^2exp(x^3)

14. Graph $f(x) = \sin x$ for $0 < x < 2\pi$. Find the area under the sine curve for $0 < x < 2\pi$. What is the relationship between this area and $\int_0^{2\pi} \sin x \, dx$?

Solution: Fill in the values between 0 and 6.3 at intervals of 0.1 in column A. Then enter the formula, =SIN(A1), into cell B1. Then perform a click-and-drag to compute the function values in column B. Create the graph using the Chart Wizard's XY (Scatter) plot. The result is as follows:

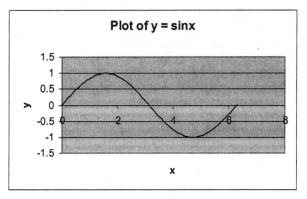

Plot of y = sinx

For $\int_0^{2\pi} \sin x \, dx$, both $F(x) = -\cos x$ and $\int_0^{2\pi} \sin x \, dx = F(2\pi) - F(0) = -\cos(2\pi) + \cos(0) = -1 + 1 = 0$.

However, note that $\int_0^{\pi} \sin x \, dx = F(\pi) - F(0) = -\cos(\pi) + \cos(0) = -(-1) + 1 = 2$ and $\int_{\pi}^{2\pi} \sin x \, dx = F(2\pi)$

$- F(\pi) = -1 - 1 = -2$. The area under the sine curve from $x = 0$ to $x = 2\pi$ is $2 + 2 = 4$ square units. When the integral is evaluated from $x = 0$ to $x = 2\pi$, we obtain 0, but the actual area is 4.

Application

15. Fax Inc. sells fax machines and had annual revenue (in billions of dollars) during the period 1995-2000 of approximately $r(t) = 0.15e^{0.5t}$ $(0 < t < 6)$ where t is the number of years since January 1995. Use this model to estimate the total revenue earned by Fax Inc. from January 1995 to December 2000. Show the total revenue earned graphically.

Solution: The total revenue earned is $\int_0^6 0.15e^{0.5t} dt$. The antiderivative is $F(t) = \dfrac{0.15}{0.5}e^{0.5t}$ and the total revenue earned is $F(6) - F(0) = 0.3e^3 - 0.3e^0 = 5.73$ billion. Fill in the values 0 through 6 at intervals of 0.5 in column A. Then enter the formula, =0.15*EXP(0.5*A1), into cell B1 and the function values in column B are returned with a click-and-drag. Enter the point (6, 0) by using cells A14 and B14 to tie down the function:

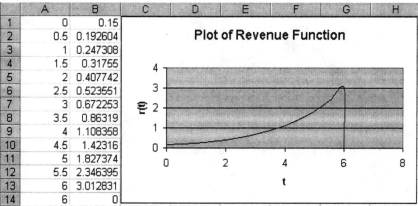

	A	B	C	D	E	F	G	H
1	0	0.15						
2	0.5	0.192604			Plot of Revenue Function			
3	1	0.247308						
4	1.5	0.31755						
5	2	0.407742						
6	2.5	0.523551						
7	3	0.672253						
8	3.5	0.86319						
9	4	1.108358						
10	4.5	1.42316						
11	5	1.827374						
12	5.5	2.346395						
13	6	3.012831						
14	6	0						

The area bounded by the function $f(t)$, the line $t = 6$, the t-axis, and the $f(t)$ axis represents the total revenue earned.

Chapter Review Exercises

1. If $f(x) = e^x + \dfrac{5}{x} - 4x^2$, find $F(x)$. Use the limit definition to evaluate the derivative of $F(x)$ at $x = 2$ and show that you get $f(2)$ for your answer.

2. Use the method of substitution to evaluate $\int x(3x^2 + 3)^3\, dx$ and then use the limit definition to evaluate the derivative of $F(x)$ at $x = 1$ and show that you get $f(1)$ for your answer.

3. Approximate the value of $\displaystyle\int_0^1 \dfrac{x}{1+x^2}\, dx$ by using Riemann sums with $n = 10$, $n = 100$, and $n = 1{,}000$ and compare these with the answer obtained using the fundamental theorem of calculus.

4. The distribution of male heights is given by $f(x) = \dfrac{1}{\sqrt{2\pi}(2.5)} e^{-\dfrac{(x-70)^2}{2(2.5)^2}}$. Use a Riemann sum with $n = 10$ to approximate the percent of males with heights between 60 and 70 inches. Also graph the function.

5. Graph $f(x) = \cos x$ for $0 < x < 2\pi$. Find the area under the cosine curve for $0 < x < 2\pi$. What is the relationship between this area and $\displaystyle\int_0^{2\pi} \cos x\, dx$?

CHAPTER 7

Further Integration Techniques and Applications of the Integral

7-1 Integration by Parts

1. Find the value of $\displaystyle\int_0^2 \frac{xe^x}{(x+1)^2}\,dx$ using integration by parts and compare your answer to a right Riemann sum, a left Riemann sum, and a midpoint Riemann sum using $n = 10$.

Solution: Applying integration by parts, we obtain

$$\int \frac{xe^x}{(1+x)^2}\,dx = \frac{e^x}{x+1}$$

Therefore,

$$\int_0^2 \frac{xe^x}{(x+1)^2}\,dx = F(2) - F(0), \text{ where } F(x) = \frac{e^x}{x+1}.$$

$$\int_0^2 \frac{xe^x}{(x+1)^2}\,dx = \frac{e^2}{2+1} - \frac{e^0}{0+1} \approx 2.463 - 1 \approx 1.463.$$

For the Riemann sums, $\Delta x = 2/10 = 0.2$. If $f(x) = \dfrac{xe^x}{(x+1)^2}$, the left, right, and midpoint Riemann sums

are $[f(0) + f(0.2) + f(0.4) + f(0.6) + f(0.8) + f(1.0) + f(1.2) + f(1.4) + f(1.6) + f(1.8)](0.2)$,

$[f(0.2) + f(0.4) + f(0.6) + f(0.8) + f(1.0) + f(1.2) + f(1.4) + f(1.6) + f(1.8) + f(2.0)](0.2)$, and

$[f(0.1) + f(0.3) + f(0.5) + f(0.7) + f(0.9) + f(1.1) + f(1.3) + f(1.5) + f(1.7) + f(1.9)](0.2)$.

In the Excel spreadsheet, fill in 0 through 1.8 at intervals equal to 0.2 in column A, and enter the formula, =(A1*EXP(A1))/(A1+1)^2, into cell B1. Then copy cell B1 to cells B2 through B10. Finally, enter the formula, =SUM(B1:B10)*0.2, into cell C1. The left Riemann sum is 1.300063.

Next, fill in the numbers 0.2 through 2.0 at intervals equal to 0.2 in column D, and enter the formula, =(D1*EXP(D1))/(D1+1)^2, into cell E1. Then copy cell E1 to cells E2 through E10. Finally, enter the formula, =SUM(E1:E10)*0.2, into cell F1. The right Riemann sum is 1.628465.

Next, fill in the midpoints of the intervals into column G and enter the formula, =(G1*EXP(G1))/(G1+1)^2, into cell H1. After forming the function values in column H, enter the formula, =SUM(H1:H10)*0.2, into cell I1. The midpoint Riemann sum is 1.46239. This value is very close to the exact value obtained from using the fundamental theorem of calculus.

	A	B	C	D	E	F	G	H	I
1	0	0	1.300063	0.2	0.169639	1.628465	0.1	0.091336	1.46239
2	0.2	0.169639		0.4	0.304454		0.3	0.23962	
3	0.4	0.304454		0.6	0.427059		0.5	0.366383	
4	0.6	0.427059		0.8	0.549516		0.7	0.48776	
5	0.8	0.549516		1	0.67957		0.9	0.613197	
6	1	0.67957		1.2	0.823169		1.1	0.749338	
7	1.2	0.823169		1.4	0.985639		1.3	0.901718	
8	1.4	0.985639		1.6	1.172315		1.5	1.075605	
9	1.6	1.172315		1.8	1.38895		1.7	1.276504	
10	1.8	1.38895		2	1.642012		1.9	1.510487	

If n is increased to $n = 1,000$ and the above technique is repeated, the left Riemann sum is found to be 1.461377, the right Riemann sum is found to be 1.464661, and the midpoint Riemann sum is found to equal 1.463019. All three are close to the true value, 1.463.

2. The demand for a software package over the next 10 years is approximated by the following equation:

$$D(t) = 500(20 + te^{-0.1t}), \; 0 < t < 10.$$

Determine the total demand for the software over the next 10 years and represent this total demand as an area under the demand curve.

Solution: We obtain the following from an integration by parts: $\int te^{-0.1t} dt = -10te^{-0.1t} - 100e^{-0.1t}$. The total demand over the next 10 years is $\int_0^{10} D(t)dt = \int_0^{10} 10,000 dt + 500 \int_0^{10} te^{-0.1t} dt$. Evaluating the two integrals on the right side we get $100,000 + 500[-10(10)e^{-1} - 100e^{-1} + 100e^0] \approx 113,212$ units sold. To plot the demand curve, enter the values from 0 to 10 by intervals of 0.5. Then enter the formula, =500*(20+A1*EXP(-0.1*A1)), into cell B1 to generate the other function values shown in column B. The point (10, 0) is used to tie the curve down on the right side. The total demand is the area bounded by the demand function $D(t)$, the line $t = 10$, the t-axis, and the $D(t)$-axis. The graph of $D(t)$ indicates that we expect the demand to increase slowly over the 10-year period and be above 10,000 units in value each year:

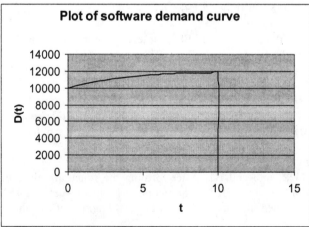

7-2 Area Between Two Curves and Applications

The Area Between Two Graphs

3. Plot the functions $f(x) = x$ and $g(x) = x^2$ for $-2 < x < 2$. Then find the area between the two functions between the intersection points of $f(x) = x$ and $g(x) = x^2$. To find the intersection points of the two functions, first find their points of intersection by setting $x = x^2$. This is equivalent to $x^2 - x = 0$ or $x(x - 1) = 0$. The solutions are $x = 0$ and $x = 1$. The points of intersection are $(0, 0)$ and $(1, 1)$. Finally, find the area under each, subtract the areas, and show that you get the same as you do by finding

$$\int_0^1 (x - x^2)\,dx\,.$$

Solution: The following plot obtained using the Chart Wizard shows the two functions plotted on the same coordinate system. The area we are to find is the small sliver between the points $(0, 0)$ and $(1, 1)$.

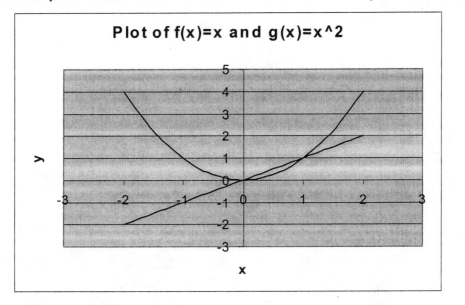

The area is $\displaystyle\int_0^1 (x - x^2)\,dx = F(1) - F(0) = 0.167$, where $F(x) = \dfrac{x^2}{2} - \dfrac{x^3}{3}$. The area under $f(x) = x$ for x

between 0 and 1 is $\displaystyle\int_0^1 x\,dx = 0.5$, and the area under $g(x) = x^2$ for x between 0 and 1 is $\displaystyle\int_0^1 x^2\,dx = 0.333$.

The difference between these two areas is 0.167.

4. Find the area of the region bounded by the graph of $y = x^2 - 3x + 2$ and the x-axis.

Solution: First, graph the quadratic function in the usual manner using Excel's Chart Wizard. The graph is as follows:

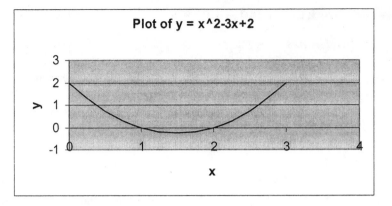

Plot of y = x^2-3x+2

The equation of the x-axis is $y = 0$. Take $f(x) = 0$ and $g(x) = x^2 - 3x + 2$. The area between the curves is

$$\int_1^2 (f(x) - g(x))dx = \int_1^2 -x^2 + 3x - 2 \ dx = F(2) - F(1), \text{ where } F(x) = -\frac{x^3}{3} + \frac{3}{2}x^2 - 2x. \ F(2) - F(1) =$$

$-0.667 + 0.833 = 0.166.$

5. Find the area of the region bounded by the graphs of $f(x) = x^3 - x^2 - 2x$ and $g(x) = -x^2 + 2x$.

Solution: First, let's construct a graph of the curves using the Chart Wizard:

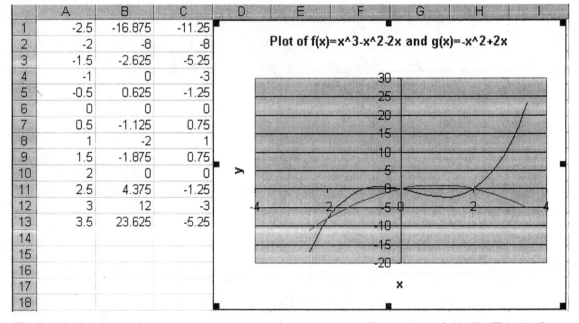

	A	B	C
1	-2.5	-16.875	-11.25
2	-2	-8	-8
3	-1.5	-2.625	-5.25
4	-1	0	-3
5	-0.5	0.625	-1.25
6	0	0	0
7	0.5	-1.125	0.75
8	1	-2	1
9	1.5	-1.875	0.75
10	2	0	0
11	2.5	4.375	-1.25
12	3	12	-3
13	3.5	23.625	-5.25
14			
15			
16			
17			
18			

Plot of f(x)=x^3-x^2-2x and g(x)=-x^2+2x

The Excel plot shows the curves intersecting at the points $(-2, -8)$, $(0, 0)$, and $(2, 0)$. This is also verifiable algebraically. If we set the functions equal to each other and solve for x, we obtain the equation, $x^3 - x^2 - 2x = -x^2 + 2x$ or $x^3 - 4x = 0$, which factors as $x(x - 2)(x + 2) = 0$. Therefore, the solutions are $x = -2$, 0, and 2. If we substitute these values for x back into either $f(x)$ or $g(x)$, we obtain the intersection points. From the graph, we see that $f(x) > g(x)$ when $-2 < x < 0$ and $f(x) < g(x)$ when $0 < x < 2$.

From this discussion, we see that the area of the region bounded by the graphs of $f(x) = x^3 - x^2 - 2x$ and $g(x) = -x^2 + 2x$ is expressed as $A = \int_{-2}^{0} (f(x) - g(x))\, dx + \int_{0}^{2} (g(x) - f(x))\, dx$. Substituting for $f(x)$ and $g(x)$, we obtain $A = \int_{-2}^{0} x^3 - 4x\, dx + \int_{0}^{2} -x^3 + 4x\, dx = \left[\dfrac{x^4}{4} - 2x^2 \right]_{-2}^{0} + \left[-\dfrac{x^4}{4} + 2x^2 \right]_{0}^{2} = 4 + 4 = 8$ square units.

Consumers' and Producers' Surplus

6. The demand equation for several colleges and universities is $q = 9859.4 - 2.2p$, where q is the enrollment at a college or university and p is the average annual tuition it charges. Officials at Midwestern University have developed a policy whereby the number of students it will accept per year at tuition level p dollars is given by the supply equation $q = 100 + 0.5p$. Find the equilibrium tuition price and the consumers' and producers' surplus at this tuition level. What is the total social gain at the equilibrium tuition? Draw graphs to help illustrate the various concepts.

Solution: The demand equation as a function of supply is $D(q) = 4481.55 - 0.45q$ and the supply equation as a function of q is $S(q) = 2q - 200$. The equilibrium price is the price when supply = demand or $9859.4 - 2.2p = 100 + 0.5p$. Solving for p, we find $\bar{p} = \$3614.6$. This is the equilibrium tuition price. The enrollment for this price is $\bar{q} = 1907$. Therefore, the consumers' surplus at this price is:

$$CS = \int_{0}^{\bar{q}} (D(q) - \bar{p})\, dq = \int_{0}^{1907} (4481.55 - 0.45q - 3614.6)\, dq = \int_{0}^{1907} (866.95 - 0.45q)\, dq = F(1907) - F(0),$$ where

$F(q) = 866.95q - 0.23q^2$. Therefore, $CS = 816{,}844.38 - 0 = 816{,}844$. The producers' surplus at the equilibrium price is: $PS = \int_{0}^{\bar{q}} (\bar{p} - S(q))\, dq = \int_{0}^{1907} (3614.6 - 2q + 200)\, dq = \int_{0}^{1907} (3814.6 - 2q)\, dq = F(1907) - F(0)$, where $F(q) = 3814.6q - q^2$. Therefore $PS = 3{,}637{,}793$. The total social gain is $CS + PS = 4{,}454{,}637$.

The first of the next two Excel spreadsheets gives a graphical explanation of the consumers' surplus. The graph shows the functions $p = D(q)$ and $p = \bar{p}$. Enter the formula, =4482-.45*A2, into cell B2, and a click-and-drag returns the required points for plotting. In a similar fashion, the second of the next two spreadsheets gives a graphical explanation of the producers' surplus, using the formula, =2*A2-200. The graph shows the functions $p = S(q)$ and $p = \bar{p}$. Add the arrow and textbox to the two Excel graphs by using the Microsoft drawing tool.

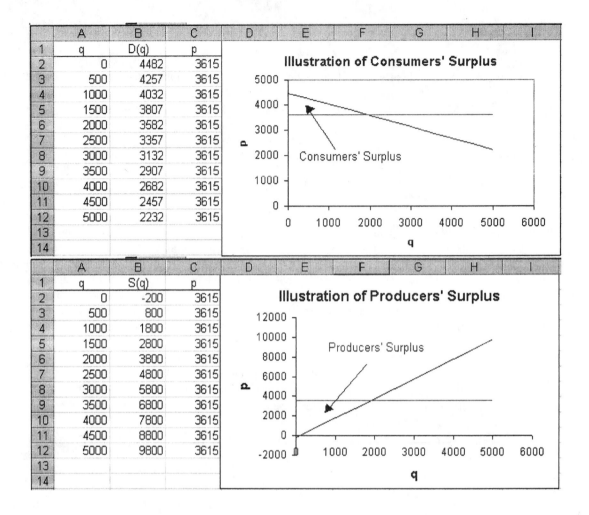

	A	B	C	D	E	F	G	H	I
1	q	D(q)	p						
2	0	4482	3615						
3	500	4257	3615						
4	1000	4032	3615						
5	1500	3807	3615						
6	2000	3582	3615						
7	2500	3357	3615						
8	3000	3132	3615						
9	3500	2907	3615						
10	4000	2682	3615						
11	4500	2457	3615						
12	5000	2232	3615						
13									
14									

	A	B	C	D	E	F	G	H	I
1	q	S(q)	p						
2	0	-200	3615						
3	500	800	3615						
4	1000	1800	3615						
5	1500	2800	3615						
6	2000	3800	3615						
7	2500	4800	3615						
8	3000	5800	3615						
9	3500	6800	3615						
10	4000	7800	3615						
11	4500	8800	3615						
12	5000	9800	3615						
13									
14									

7-3 Averages and Moving Averages

Averages

7. The price of a newly introduced product decreases according to the model $C = 15 - 0.009t^2$, $0 \le t \le$ 12 where t is the time from the introduction of the product in months. Use a definite integral to find the average cost over the year period. Also assume that one item is purchased per month beginning with month $t = 0$ and find the average cost of theses 13 items. Compare this result with the answer obtained with the integral.

Solution: The average cost is $= \dfrac{1}{12} \displaystyle\int_{0}^{12} 15 - 0.009t^2 \ dt = \dfrac{1}{12}\left[15t - 0.003t^3\right]_{t=0}^{t=12} = \$14.57$. To find the

average cost of 13 items purchased one per month, fill in the numbers 0 through 12 in column A, and enter the formula, =15-0.009*A1^2, into cell B1. Then compute the monthly costs by using a click-and-drag. The formula, =AVERAGE(B1:B13), entered into cell C1 will compute the average, $14.55. Note that this is very close to that found by integration.

	A	B	C
1	0	15	14.55
2	1	14.991	
3	2	14.964	
4	3	14.919	
5	4	14.856	
6	5	14.775	
7	6	14.676	
8	7	14.559	
9	8	14.424	
10	9	14.271	
11	10	14.1	
12	11	13.911	
13	12	13.704	

8. The average value of a function $f(x)$ on an interval [a, b] is $\bar{f} = \dfrac{1}{b-a}\displaystyle\int_a^b f(x)dx$. **In general, the**

average $\bar{f}$ of a positive function over the interval [a, b] gives the height of the rectangle over the interval [a, b] having the same area as the area under the graph of $f(x)$). For the function,

$f(x) = \begin{cases} -x, & \text{if } -1 < x < 0 \\ x, & \text{if } 0 < x < 1 \end{cases}$, find the average value of the function. Also construct this function

and the function, $y = \bar{f}$, $-1 < x < 1$ on the same coordinate system and verify the relationship between the two areas.

Solution: The average value of $f(x)$ is $\bar{f} = \dfrac{1}{2}\displaystyle\int_{-1}^{0} -x\,dx + \dfrac{1}{2}\displaystyle\int_{0}^{1} x\,dx$ = 0.5. In the following graph, the

dotted line is the line $y = \bar{f}$ and the solid line is the graph of $f(x)$:

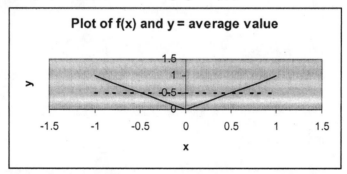

Plot of f(x) and y = average value

The area under the dotted line for $-1 < x < 1$ is (0.5)(2) = 1. The graph of $f(x)$ for $-1 < x < 1$ is composed of two triangles each having base = 1 and height = 1. Therefore the area under the graph of $f(x)$ is twice $\dfrac{1}{2}bh$ or twice $\dfrac{1}{2}(1)(1) = \dfrac{1}{2}$, and twice this is 1.

Moving Averages

9. The following table gives the closing stock prices for Computers Inc. for twenty consecutive days.

Day	Price	Day	Price
1	130	11	143
2	134	12	141
3	127	13	150
4	136	14	154
5	129	15	149
6	134	16	148
7	133	17	155
8	137	18	150
9	140	19	158
10	138	20	162

Plot the prices and the four-day moving averages and note the smoothing that occurs for the moving averages.

Solution: Fill in the 20 days in column A, and enter the stock prices into column B. Then enter the formula, =AVERAGE(B1:B4), into cell C4. A click-and-drag produces the moving average in column C, and the Chart Wizard provides the following plot:

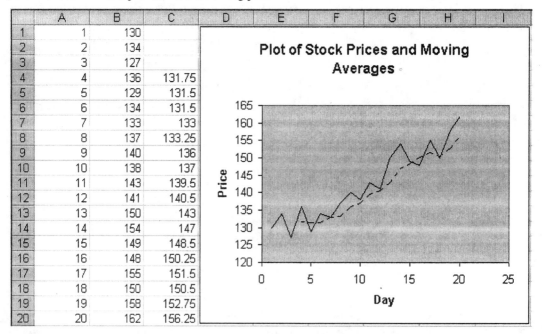

Note that the moving averages have been plotted as a dashed line.

10. Find the 2-unit moving average of $f(t) = t^3$ and plot both.

Solution: The 2-unit moving average is $= \bar{f}(x) = \dfrac{1}{2}\displaystyle\int_{x-2}^{x} t^3\, dt = \dfrac{1}{8}\left[t^4\right]_{x-2}^{x} = x^3 - 3x^2 + 4x - 2$. To graph $f(x)$

and $\bar{f}(x)$, fill in the numbers from -2 to 2 in increments of 0.2 in column A. Enter the formula, =A1^3, into cell B1 and the formula, =A1^3-3*A1^2+4*A1-2, into cell C1. Obtain the function values and use the Chart Wizard to plot the two graphs. The function, $\bar{f}(x)$, is shown as a dashed line and the function, $f(x)$, is shown as a solid line.

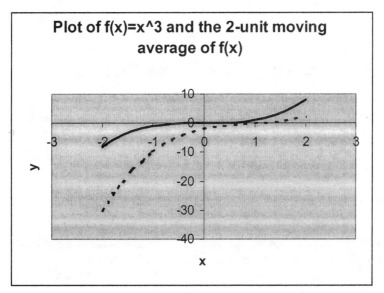

Plot of f(x)=x^3 and the 2-unit moving average of f(x)

7-4 Continuous Income Streams

Total Revenue of a Continuous Income Stream

11. The revenue (in thousands of dollars) per day for a new Internet business is given by $R(t) = t$, for $0 \le t \le 10$. Approximate the total revenue by using the right Riemann sum with $n = 10$. The expression is $[R(1) + R(2) + \cdots R(10)](1)$. This Riemann sum assumes $R(1) = 1$ thousand per day for day 1, $R(2) = 2$ thousand for day 2, etc. Find the total revenue for the 10-day period using a definite integral and compare your answers. If this revenue function were valid for the full year, what would the total revenue equal?

Solution: Fill in the values 1 through 10 in column A and enter the formula, =SUM(A1:A10), into cell B1. Excel returns the value 55 as shown.

	A	B
1	1	55
2	2	
3	3	
4	4	
5	5	
6	6	
7	7	
8	8	
9	9	
10	10	

This value could also be computed by using the formula for the sum of the first n positive integers, $\frac{n(n+1)}{2} = \frac{10(11)}{2} = 55$. The integral representation of the sum is $\int_0^{10} t \; dt = \left[\frac{t^2}{2}\right]_0^{10} = 50$. The Riemann approximation gives \$55,000 and the definite integral gives \$50,000. If the revenue function were valid for the full year, the total revenue would be $\int_0^{365} t \; dt = \left[\frac{t^2}{2}\right]_0^{365} = 66{,}612.5$.

Future Value of a Continuous Income Stream

12. If $R(t) = \$30{,}000$ per year, $0 \le t \le 10$ (years), at 7%, find total value of the given income stream and its future value at the end of the given interval, assuming continuous compounding. Also, find the left Riemann sum approximation to the future value using $n = 10$, 100, and 1000 subdivisions.

Solution: The total value of the income stream is $\int_0^{10} R(t)dt = \int_0^{10} 30000dt = \$300{,}000$. The future value at the end of 10 years is $e^{rb} \int_a^b R(t)e^{-rt}dt = e^{0.07(10)} \int_0^{10} 30{,}000 e^{-0.07t} dt = \$434{,}465.45$. The Riemann sum approximation to the future value using 10 subdivisions is \$449,849.10 as shown below.

C1		=	=EXP(0.7)*30000*EXP(-0.07*A1)		
	A	B	C	D	E	F
1	0	30000	60412.58	449849.1		
2	1	30000	56328.32			
3	2	30000	52520.18			
4	3	30000	48969.49			
5	4	30000	45658.85			
6	5	30000	42572.03			
7	6	30000	39693.89			
8	7	30000	37010.34			
9	8	30000	34508.21			
10	9	30000	32175.25			

If 100 subdivisions are used, the Riemann sum is $435,987.80. If 1000 subdivisions are used, the Riemann sum is $434,617.50.

Present Value of a Continuous Income Stream

13. If $R(t) = \$30,000$ per year, $0 \le t \le 10$ (years), at 7%, find its present value at the beginning of the given interval, assuming continuous compounding. Also, find the midpoint Riemann sum approximation by using $n = 10$, 100, and 1000 subdivisions.

Solution: The present value of this continuous income stream is

$$e^{ra} \int_a^b R(t)e^{-rt}dt = e^{0.07(0)} \int_0^{10} 30{,}000e^{-0.07t}dt = \$215{,}749.16$$

The midpoint Riemann sum approximation to the present value using 10 subdivisions is $215,705.10 as shown below.

	B1		=	=30000*EXP(-0.07*A1)	
	A	B	C	D	E
1	0.5	28968.16	215705.1		
2	1.5	27009.74			
3	2.5	25183.71			
4	3.5	23481.14			
5	4.5	21893.67			
6	5.5	20413.52			
7	6.5	19033.44			
8	7.5	17746.66			
9	8.5	16546.88			
10	9.5	15428.21			
11					

If 100 subdivisions are used, the Riemann sum is $215,748.70. If 1000 subdivisions are used, the Riemann sum is $215,749.20. Note how close the midpoint Riemann sum approximation with 1000 subdivisions is to the answer given by the definite integral for the present value.

7-5 Improper Integrals and Applications

Improper Integral with an Infinite Limit of Integration

14. Evaluate $\int_1^{\infty} \frac{1}{x^3}dx$. In addition, plot the graph of the integrand and approximate the value of the integral by the use of a Riemann sum.

Solution: The integral is evaluated as follows. $\lim_{M \to \infty} \int_1^M \frac{1}{x^3}dx = \lim_{M \to \infty}\left(-\frac{1}{2M^2} + 0.5\right) = 0.5$. To plot the function $f(x) = \frac{1}{x^3}$, enter the number 1 into cell A1. The pull-down sequence, Edit $\to$ Fill $\to$ Series, produces the following dialog box:

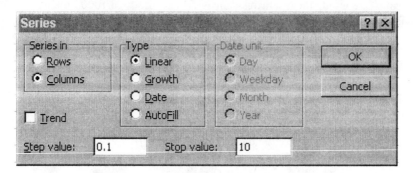

The selections shown above will fill the numbers 1 through 10 in steps of 0.1 in column A. Now enter the formula, =1/A1^3, into cell B1 and perform a click-and-drag to fill out the column. Then use the Chart Wizard to construct the following graph:

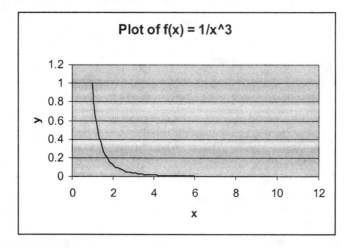

The formula, =SUM(B1:B90)*0.1, in cell C1 returns 0.547, the left Riemann sum estimate for the area under $f(x)$ for $1 < x < 10$. The function value at $x = 10$ is 0.001 and we see that there will be very little area under the curve to the right of $x = 10$. As the value of n increases, the Riemann sums will get closer and closer to 0.5.

15. Evaluate $\displaystyle\int_{-\infty}^{0} \frac{dx}{(1-2x)^{3/2}}$. In addition, plot the graph of the integrand and approximate the value of the integral by the use of a Riemann sum.

Solution: The integral is evaluated as $\displaystyle\lim_{M\to-\infty} \int_{M}^{0} \frac{dx}{(1-2x)^{3/2}} = \lim_{M\to-\infty}\left(1 - \frac{1}{\sqrt{1-2M}}\right) = 1$. To plot the function, $f(x) = \dfrac{1}{(1-2x)^{3/2}}$, start by entering the number -10 into cell A1. The pull-down sequence,

<u>E</u>dit → F<u>i</u>ll → <u>S</u>eries, produces the following dialog box:

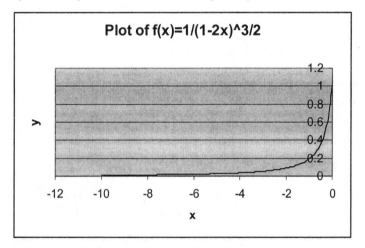

The selections shown will fill in the required values for column A. Enter the formula, =1/(1-2*A1)^1.5, into cell B1 and compute the required function values for plotting with the Chart Wizard:

Plot of f(x)=1/(1-2x)^3/2

Finally, when the formula, =SUM(B1:B100)*0.1, is entered into cell C1, Excel returns the area, 0.735. We would need to decrease the step value and go further out than $x = -10$ to get closer to the true area which equals 1.

Integrals in Which the Integrand Becomes Infinite

16. Evaluate the improper integral $\int_1^2 \dfrac{dx}{\sqrt[3]{x-1}}$. Plot the integrand for x between 1 and 2. Approximate the integral by use of a Riemann sum.

Solution: The integral is defined as $\lim\limits_{r\to 1^+} \int_r^2 (x-1)^{-1/3}\, dx = \lim\limits_{r\to 1^+}\left[\dfrac{3}{2} - \dfrac{3}{2}(r-1)^{2/3}\right] = \dfrac{3}{2}$. Start by filling in the values 1.1 through 2 in increments of 0.1 in column A. Enter the formula, =(A1-1)^(-1/3), into cell B1 and the function values in column B are formed. Finally, enter the formula, =SUM(B1:B10)*0.1, in cell C1. Excel returns the value, 1.34, of the right Riemann sum with $n = 10$. This is an approximation to the true value 1.5. The plot of the segment is as follows:

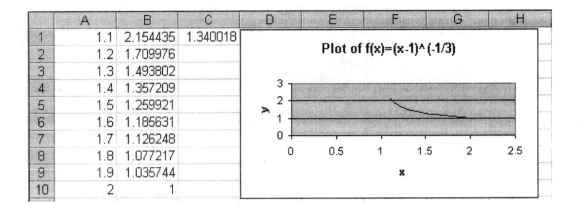

7-6 Differential Equations and Applications

Simple Differential Equations

17. Solve the following simple differential equation and graph some of its solution curves:

$$\frac{dy}{dx} = e^x + \cos x - x^2$$

Find the particular solution curve determined by the condition $y = 0$ when $x = 0$.

Solution: The general solution is $y = \int e^x + \cos x - x^2 \, dx = e^x + \sin x - \dfrac{x^3}{3} + C$. Suppose we decide to graph the solution curves corresponding to $C = -5, -3, -1, 0, 1, 3,$ and 5. To graph the solution curves, enter values from -2 to 2 in column A. Then enter the formula, =EXP(A1)+SIN(A1)-A1^3/3-5, into cell B1, the formula, =B1+2, into cell C1, and so forth until we enter the formula, =B1+12, into cell H1. Excel returns the following values on the solutions curves:

	A	B	C	D	E	F	G	H
1	-2	-3.1073	-1.1073	0.892705	1.892705	2.892705	4.892705	6.892705
2	-1.5	-4.64936	-2.64936	-0.64936	0.350635	1.350635	3.350635	5.350635
3	-1	-5.14026	-3.14026	-1.14026	-0.14026	0.859742	2.859742	4.859742
4	-0.5	-4.83123	-2.83123	-0.83123	0.168772	1.168772	3.168772	5.168772
5	0	-4	-2	0	1	2	4	6
6	0.5	-2.91352	-0.91352	1.08648	2.08648	3.08648	5.08648	7.08648
7	1	-1.77358	0.226419	2.226419	3.226419	4.226419	6.226419	8.226419
8	1.5	-0.64582	1.354184	3.354184	4.354184	5.354184	7.354184	9.354184
9	2	0.631687	2.631687	4.631687	5.631687	6.631687	8.631687	10.63169

Using the Chart Wizard, we obtain the following solution curves for the differential equation, $\dfrac{dy}{dx} = e^x + \cos x - x^2$. Next we obtain the particular solution to the equation, $y = 0$ when $x = 0$. The

100

requirement that $y = 0$ when $x = 0$ is substituted in the general solution, $y = e^x + \sin x - \dfrac{x^3}{3} + C$, to yield

$0 = e^0 + \sin 0 - \dfrac{0}{3} + C = 1 + 0 - 0 + C$. Solving for C, we get $C = -1$. This particular solution is shown as the dashed curve in the set of solution curves. Some values on this particular solution are seen in column D of the above Excel spreadsheet.

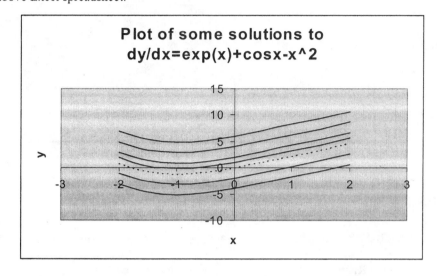

Plot of some solutions to dy/dx=exp(x)+cosx-x^2

Separable Differential Equations

18. Solve the following separable differential equation: $2x\dfrac{dy}{dx} = 3y$.

Find the particular solution curve determined by the condition, $y = 4$ when $x = 4$, and graph this particular solution.

Solution: Separating the variables, we have $2x\,dy = 3y\,dx$. Then we can write $\dfrac{2dy}{y} = \dfrac{3dx}{x}$. Integrating both sides, we get $2 \ln y = 3 \ln x + C$. We can also write this as $\ln y^2 = \ln x^3 + C$, or $\ln \dfrac{y^2}{x^3} = C$, or

$\dfrac{y^2}{x^3} = e^C = K$, or finally, $y^2 = Kx^3$. There are no solutions for which $x < 0$ because y^2 is positive, and if

$x < 0$, Kx^3 is negative and y^2 cannot be negative. When $x > 0$, the solution can be written as $y = \pm \sqrt{Kx^3}$. The particular solution determined by the condition $y = 4$ when $x = 4$ occurs when $4^2 = K4^3$ or $K = 0.25$. The particular solution, therefore, is $y = \pm\sqrt{0.25x^3} = \pm\ 0.5\sqrt{x^3}$. Graph this particular solution by first filling in the numbers 0 through 5 in column A. Enter the formula, =0.5*SQRT(A1^3), into cell B1 and the formula, =-0.5*SQRT(A1^3), into cell C1. Excel returns the required values. Now use the Chart Wizard to construct the particular solution curve:

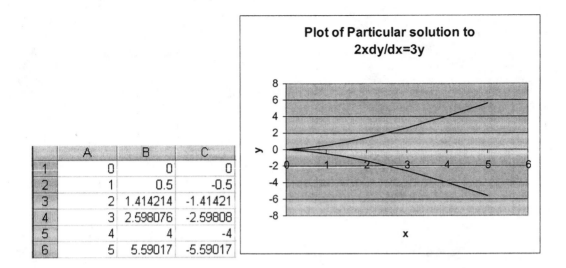

	A	B	C
1	0	0	0
2	1	0.5	-0.5
3	2	1.414214	-1.41421
4	3	2.598076	-2.59808
5	4	4	-4
6	5	5.59017	-5.59017

Chapter Review Exercises

1. Find the value of $\int_{0}^{1} (x+1)e^x \, dx$ using integration by parts and compare your answer to a right Riemann sum, a left Riemann sum, and a midpoint Riemann sum using $n = 10$.

2. Graph the functions $f(x) = x$ and $g(x) = x^4$ and find the area enclosed by the two graphs.

3. The average value of a function $f(x)$ on an interval $[a, b]$ is $\bar{f} = \dfrac{1}{b-a}\int_{a}^{b} f(x)dx$. For the function, $f(x) = e^x$, find the average value of the function over $[-1, 1]$. Also construct this function and the function $y = \bar{f}$, $-1 < x < 1$ on the same coordinate system and verify the relationship between the two areas.

4. The following table gives the closing stock prices for Books.com for twenty consecutive days.

Day	Price	Day	Price
1	60	11	75
2	70	12	75
3	65	13	80
4	55	14	85
5	60	15	90
6	60	16	85
7	80	17	90
8	85	18	100
9	80	19	100
10	80	20	100

Plot the prices and the four-day moving averages and note the smoothing that occurs for the moving averages.

5. The revenue (in thousands of dollars) per day for a new Internet business is given by $R(t) = 2t + 1$, for $0 \leq t \leq 20$. Approximate the total revenue by using the right Riemann sum with $n = 10$. Find the total revenue for the 20-day period using a definite integral and compare your answers. If this revenue function were valid for the full year, what would the total revenue equal?

6. If $R(t) = \$10,000$ per year, $0 \leq t \leq 5$ (years), at 5%, find total value of the given income stream and its future value at the end of the given interval, assuming continuous compounding. Also, find the left Riemann sum approximation to the future value using $n = 10$, 100, and 1000 subdivisions.

7. If $R(t) = \$10,000$ per year, $0 \leq t \leq 5$ (years), at 5%, find its present value at the beginning of the given interval, assuming continuous compounding. Also, find the midpoint Riemann sum approximation using $n = 10$, 100, and 1000 subdivisions.

8. Evaluate $\displaystyle\int_{1}^{\infty} \frac{1}{e^x} dx$. In addition, plot the graph of the integrand and approximate the value of the integral by the use of a Riemann sum.

9. Evaluate the improper integral $\displaystyle\int_{1}^{2} \frac{4}{\sqrt{x-1}} dx$. Plot the integrand for x between 1 and 2. Approximate the integral by use of a Riemann sum.

10. Solve the following separable differential equation. Find the particular solution curve determined by the condition $y = 0$ when $x = 0$ and graph it.

$$\frac{dy}{dx} = xe^{x^2} + \sin x - x + 2$$

CHAPTER 8

Functions of Several Variables

8-1 Functions of Several Variables from the Numerical and Algebraic Viewpoints

Real Valued Functions of Two Variables

1. A sphere centered at (0, 0, 0) and having radius equal to 1 has the equation $x^2 + y^2 + z^2 = 1$. If we solve this equation for z, we obtain two functions of two variables. We can express z in terms of x and y as $z = f(x, y) = \pm\sqrt{1 - x^2 - y^2}$. The upper half of the sphere is expressed by the function $z = f(x, y) = \sqrt{1 - x^2 - y^2}$, whereas and the lower half of the sphere is expressed by the function $z = f(x, y) = -\sqrt{1 - x^2 - y^2}$. Use Excel to create some points on the upper half of the sphere where the x-value of the point on the sphere is 0.

Solution: First note that when choosing the values of x and y, $x^2 + y^2$ must be equal to or less than 1. Otherwise, we encounter the square root of a negative number. To find these points, fill in 21 values of 0 in column A and values of y from -1 to 1 in increments of 0.1 in column B. Then enter the formula, =SQRT(1-A2^2-B2^2), to generate the values of z in column C:

	A	B	C
1	x	y	z
2	0	-1	0
3	0	-0.9	0.43589
4	0	-0.8	0.6
5	0	-0.7	0.714143
6	0	-0.6	0.8
7	0	-0.5	0.866025
8	0	-0.4	0.916515
9	0	-0.3	0.953939
10	0	-0.2	0.979796
11	0	-0.1	0.994987
12	0	0	1
13	0	0.1	0.994987
14	0	0.2	0.979796
15	0	0.3	0.953939
16	0	0.4	0.916515
17	0	0.5	0.866025
18	0	0.6	0.8
19	0	0.7	0.714143
20	0	0.8	0.6
21	0	0.9	0.43589
22	0	1	0

The 21 points shown in the Excel worksheet are only a few of the infinite number of points that comprise the unit sphere.

Linear Function

2. Plot the section of the plane, $z = x + 4y$, for x between 0 and 4 and y between 0 and 3.

Solution: Enter the x values in cells B1:F1 and the y values into cells A2:A5. Then in cell B2, enter the formula, =B$1+4*$A2, into cell B2. After a click-and-drag from cell B2:F2, followed by a click-and-drag from B2:B5, Excel returns the required values for plotting:

	A	B	C	D	E	F
1		0	1	2	3	4
2	0	B$1+4*$A2	C$1+4*$A2	D$1+4*$A2	E$1+4*$A2	F$1+4*$A2
3	1	B$1+4*$A3	C$1+4*$A3	D$1+4*$A3	E$1+4*$A3	F$1+4*$A3
4	2	B$1+4*$A4	C$1+4*$A4	D$1+4*$A4	E$1+4*$A4	F$1+4*$A4
5	3	B$1+4*$A5	C$1+4*$A5	D$1+4*$A5	E$1+4*$A5	F$1+4*$A5

	A	B	C	D	E	F
1		0	1	2	3	4
2	0	0	1	2	3	4
3	1	4	5	6	7	8
4	2	8	9	10	11	12
5	3	12	13	14	15	16

Now select the data from A1 through F5 and invoke the Chart Wizard:

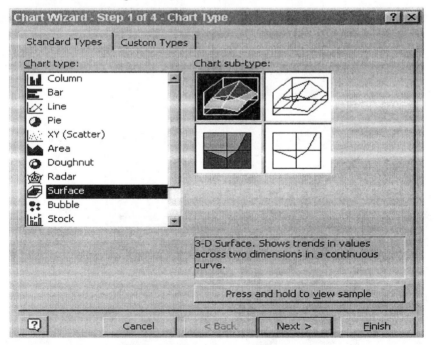

Select Surface for Chart type, followed by Chart sub-type 3-D Surface, and examine the Series tab:

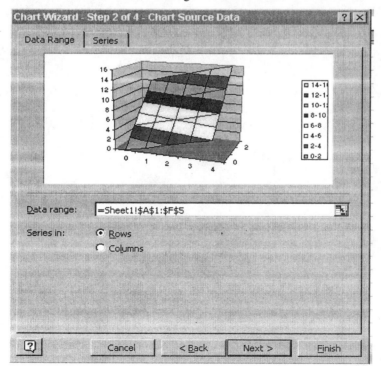

Make sure that the Category(x) axis labels shows $B$1:$F$1, since we decided to use this part of the spreadsheet as our x values. Examine the Data Range Tab and choose the Series in Rows option:

If you select Series in columns, Category (X) axis labels shows =Sheet1!$A$2:$A$5 on the Series part of Step 2, but these are the y values. After completing steps 3 and 4 of the Chart Wizard, Excel returns the required plot of the section of the plane, $z = x + 4y$:

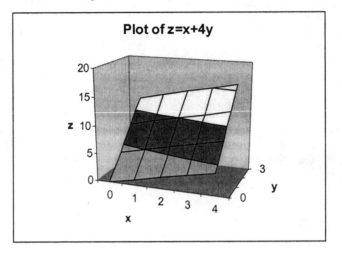

Plot of z=x+4y

Interaction Function

3. Plot the interaction function, $z = x + 4y + xy$, for values of x between 0 and 4 and y between 0 and 3.

Solution: Follow the same steps as in problem 2 and obtain the following function values for plotting:

	A	B	C	D	E	F
1		0	1	2	3	4
2	0	0	1	2	3	4
3	1	4	6	8	10	12
4	2	8	11	14	17	20
5	3	12	16	20	24	28

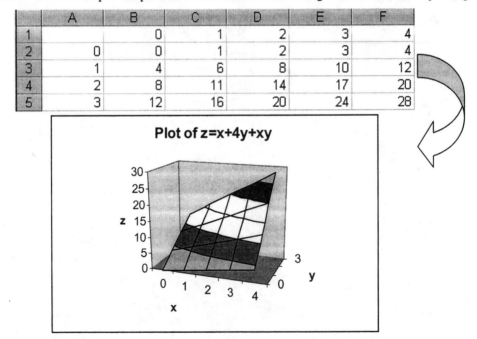

Plot of z=x+4y+xy

Notice that this surface, defined by $z = x + 4y + xy$, is not a plane, but has a curvature to it.

Equation of a Circle of Radius r Centered at the Origin

4. The equation for the circle that is centered at the origin and has a radius of 4 is $x^2 + y^2 = 16$. Plot this circle.

Solution: To plot the circle, first solve for y and obtain $y = \pm\sqrt{16 - x^2}$. The function, $f(x) = \sqrt{16 - x^2}$, describes the top half of the circle and $f(x) = -\sqrt{16 - x^2}$ describes the bottom half of the circle. To graph the circle, fill in x values from –4 to 4 in column A. Then enter the formula, =SQRT(16-A1^2), into cell B1 and the formula, =-SQRT(16-A1^2), into cell C1. A click-and-drag produces the values in columns B and C. Then use the Chart Wizard to produce the plot shown:

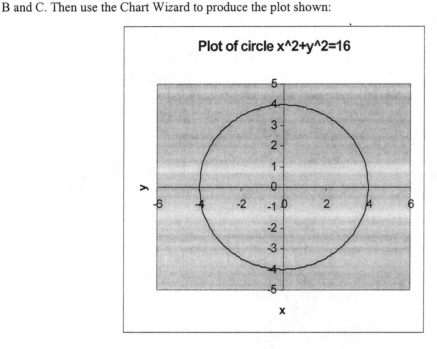

8-2 Three Dimensional Space and the Graph of a Function of Two Variables

5. An equation of the form, $z = \dfrac{x^2}{a^2} + \dfrac{y^2}{b^2}$, is called an elliptic paraboloid. Discuss the surface defined by

$z = \dfrac{x^2}{4} + \dfrac{y^2}{9}$. Generate the graph with Excel. Find the x-, y-, and z-intercepts.

Solution: First we need to compute the function values to be plotted. The grid in the xy-plane is chosen to go from –1 to 1 as shown. The x values are chosen to occupy the cells from B1 to F1. The y values occupy cells A2 to A6. Enter the formula, =B$1^2/4+$A2^2/9, into cell B1 and use click-and-drag operations to produce the function values. To help understand how this expression works, the following two panels show the formulas and the values they return to the various cells:

	A	B	C	D	E	F
1		-1	-0.5	0	0.5	1
2	-1	B$1^2/4+$A2^2/9	C$1^2/4+$A2^2/9	D$1^2/4+$A2^2/9	E$1^2/4+$A2^2/9	F$1^2/4+$A2^2/9
3	-0.5	B$1^2/4+$A3^2/9	C$1^2/4+$A3^2/9	D$1^2/4+$A3^2/9	E$1^2/4+$A3^2/9	F$1^2/4+$A3^2/9
4	0	B$1^2/4+$A4^2/9	C$1^2/4+$A4^2/9	D$1^2/4+$A4^2/9	E$1^2/4+$A4^2/9	F$1^2/4+$A4^2/9
5	0.5	B$1^2/4+$A5^2/9	C$1^2/4+$A5^2/9	D$1^2/4+$A5^2/9	E$1^2/4+$A5^2/9	F$1^2/4+$A5^2/9
6	1	B$1^2/4+$A6^2/9	C$1^2/4+$A6^2/9	D$1^2/4+$A6^2/9	E$1^2/4+$A6^2/9	F$1^2/4+$A6^2/9

	A	B	C	D	E	F
1		-1	-0.5	0	0.5	1
2	-1	0.361111	0.173611	0.111111	0.173611	0.361111
3	-0.5	0.277778	0.090278	0.027778	0.090278	0.277778
4	0	0.25	0.0625	0	0.0625	0.25
5	0.5	0.277778	0.090278	0.027778	0.090278	0.277778
6	1	0.361111	0.173611	0.111111	0.173611	0.361111

After selecting the rectangular region from A1 to F6, the four steps of the chart wizard are as follows:

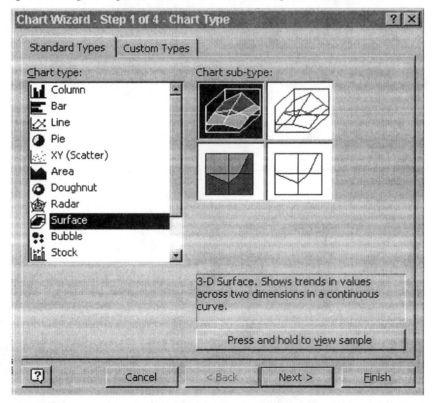

In step 1, choose Surface for Chart type and 3-D for Chart sub-type. Now examine the Data Range tab:

In step 2 for Data Range, make sure the data range is shown as A$1 to F$6. Also choose the Series in Rows option. In step 2 for Series, the Category (X) axis labels text box should show B1 to F1. This indicates that we chose B1 to F1 as our *x* values. If the Series in Columns option had been chosen instead, the Category (X) axis labels text box would show A2 to A6.

In step 3, the title, axis labels, etc can be selected. In step 4, select Finish. Excel returns the following graph:

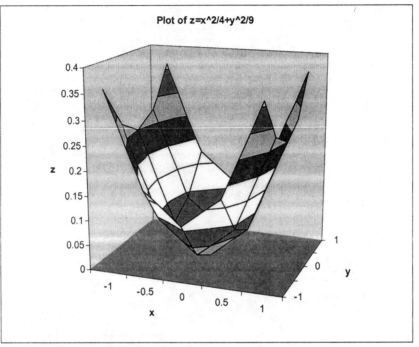

Plot of z=x^2/4+y^2/9

The surface intercepts the x, y, and z axes at $(0, 0, 0)$.

6. A sphere that is centered on the z-axis at $(0, 0, 5)$ has a radius of $r = 5$ and is defined by the equation, $x^2 + y^2 + (z - 5)^2 = 25$. Construct that portion of the bottom half of the surface of the sphere that lies above a square in the xy-plane with corners at $(1, 1)$, $(-1, 1)$, $(-1, -1)$, and $(1, -1)$.

Solution: Solving the equation of the sphere for z, we find $z = 5 \pm \sqrt{25 - x^2 - y^2}$. The bottom half of the sphere has equation $z = 5 - \sqrt{25 - x^2 - y^2}$. Since we have already shown in problem 5 of this section considerable details concerning the surface construction, this solution will supply fewer details. Enter the formula, =5-SQRT(25-B1^2-A2^2), into cell B2 and use it to construct a table of values above the square in the xy-plane with corners at $(1, 1)$, $(-1, 1)$, $(-1, -1)$, and $(1, -1)$:

	A	B	C	D	E	F	G	H	I	J
1		-1	-0.75	-0.5	-0.25	0	0.25	0.5	0.75	1
2	-1	0.204168	0.158771	0.126603	0.107404	0.101021	0.107404	0.126603	0.158771	0.204168
3	-0.75	0.158771	0.113795	0.081921	0.062896	0.05657	0.062896	0.081921	0.113795	0.158771
4	-0.5	0.126603	0.081921	0.050253	0.031348	0.025063	0.031348	0.050253	0.081921	0.126603
5	-0.25	0.107404	0.062896	0.031348	0.012516	0.006254	0.012516	0.031348	0.062896	0.107404
6	0	0.101021	0.05657	0.025063	0.006254	0	0.006254	0.025063	0.05657	0.101021
7	0.25	0.107404	0.062896	0.031348	0.012516	0.006254	0.012516	0.031348	0.062896	0.107404
8	0.5	0.126603	0.081921	0.050253	0.031348	0.025063	0.031348	0.050253	0.081921	0.126603
9	0.75	0.158771	0.113795	0.081921	0.062896	0.05657	0.062896	0.081921	0.113795	0.158771
10	1	0.204168	0.158771	0.126603	0.107404	0.101021	0.107404	0.126603	0.158771	0.204168

Finally, use the Chart Wizard to build the following plot:

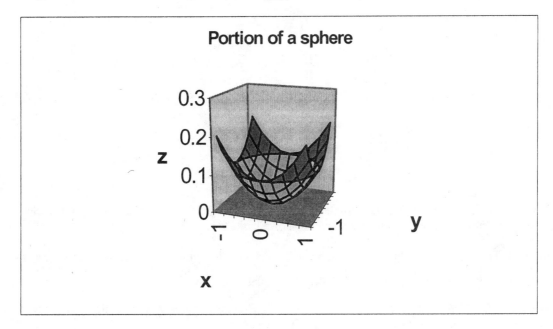

8-3 Partial Derivatives

7. The cost to manufacture x bicycles and y tricycles is $C(x,y) = 15{,}000 + 50x + 25y + 0.5xy$. The production levels of bicycles and tricycles is shown in the following table:

Production Levels			
Tricycles	Bicycles	Tricycles	Bicycles
10	5	20	5
10	10	20	10
10	15	20	15
10	20	20	20
10	25	20	25

Compute the marginal cost of manufacturing tricycles at each production level.

Solution: The marginal cost of manufacturing tricycles at a production level of x bicycles and y tricycles is $\dfrac{\partial C}{\partial y} = 25 + 0.5x$. Use columns A to hold the tricycle data and column B to hold the bicycle data.

Although we only need the bicycle data for this problem, we will need the tricycle production level data for the next problem. Now enter the formula, =25+0.5*B2, into cell C2, and use it to compute the marginal costs of manufacturing tricycles. Excel returns the marginal costs as shown in the following panel:

	A	B	C
1	y	x	25+0.5x
2	10	5	27.5
3	10	10	30
4	10	15	32.5
5	10	20	35
6	10	25	37.5
7	20	5	27.5
8	20	10	30
9	20	15	32.5
10	20	20	35
11	20	25	37.5

8. Using the data from the previous problem, compute the marginal cost of manufacturing bicycles at the production levels of bicycles and tricycles shown in the table.

Solution: The marginal cost of manufacturing bicycles at a production level of x bicycles and y tricycles is $\dfrac{\partial C}{\partial x} = 50 + 0.5y$. The tricycle data is already available in the spreadsheet. Therefore, enter the formula, =50+0.5*A2, into cell C2, and use it to compute the marginal costs of manufacturing bicycles. Excel returns the marginal costs as shown in the following panel:

	A	B	C
1	y	x	50+0.5y
2	10	5	55
3	10	10	55
4	10	15	55
5	10	20	55
6	10	25	55
7	20	5	60
8	20	10	60
9	20	15	60
10	20	20	60
11	20	25	60

8-4 Maxima and Minima

9. A sphere of radius r = 4 and centered at the origin has the equation, $x^2 + y^2 + z^2 = 16$. We can write the equation in the form of a function of two variables, $z = \pm\sqrt{16 - x^2 - y^2}$. The top half of the sphere is $z = f(x, y) = \sqrt{16 - x^2 - y^2}$. As you can visualize in your mind's eye, this surface will have a maximum at the point (0, 0, 4). Use the second-derivative test for functions of two variables to verify this. In addition, use Excel to investigate $f(x,y)$ in a neighborhood about the point (0, 0).

Solution: The first and second derivatives are

$$f_x = \frac{-x}{\sqrt{16 - x^2 - y^2}}, f_{xx} = \frac{2x^2 + y^2 - 16}{(16 - x^2 - y^2)^{3/2}}, f_y = \frac{-y}{\sqrt{16 - x^2 - y^2}}, \text{ and } f_{yy} = \frac{2y^2 + x^2 - 16}{(16 - x^2 - y^2)^{3/2}}.$$

The two first derivatives, f_x and f_y, are equal to 0 when $x = 0$ and $y = 0$. The mixed partial derivative is equal to $f_{xy} = \frac{-xy}{(16 - x^2 - y^2)^{3/2}}$. The quantity, H, is evaluated by $H = f_{xx}(0,0)f_{yy}(0,0) - [f_{xy}(0,0)]^2$

$= (-0.25)(-0.25) - 0 = 0.0625$. Since H is positive and $f_{xx}(0,0) = -0.25$ is negative, by the second-derivative test we know that f has a relative maximum at (0, 0). That is, the point (0, 0, 4) is a point in 3-dimensional space that corresponds to a relative maximum point on the surface defined by $z = f(x, y) = \sqrt{16 - x^2 - y^2}$. To investigate the region about the point (0, 0), fill in your worksheet with the values –2 to 2 in cells B1:F1 and –2 to 2 in cells A2:A6. Enter the formula, =SQRT(16-B1^2-A2^2), into cell B2 and use it to generate the required values.

	A	B	C	D	E	F
1		-2	-1	0	1	2
2	-2	2.828427	3.316625	3.464102	3.316625	2.828427
3	-1	3.316625	3.741657	3.872983	3.741657	3.316625
4	0	3.464102	3.872983	4	3.872983	3.464102
5	1	3.316625	3.741657	3.872983	3.741657	3.316625
6	2	2.828427	3.316625	3.464102	3.316625	2.828427

Note that as we move away from (0,0), the function values decrease, indicating that we are at a relative maximum. The following plot confirms that the point (0, 0, 4) is a relative maximum.

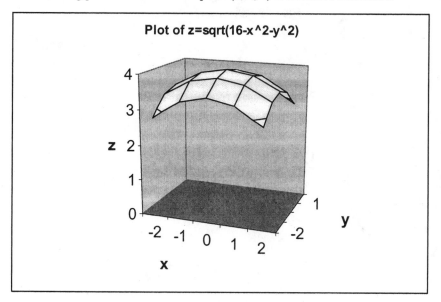

Plot of z=sqrt(16-x^2-y^2)

10. Locate the relative maxima and minima of the function, $z = x^2 - 2x + y^2 - 2y$, by using the second-derivative test. Then use Excel to investigate the neighborhood surrounding the points.

Solution: The first and second and mixed derivatives are

$$f_x = 2x - 2, f_{xx} = 2, f_y = 2y - 2, f_{yy} = 2, \text{ and } f_{xy} = 0.$$

The two first derivatives, f_x and f_y, are equal to 0 when $x = 1$ and $y = 1$. The quantity, H, is equal to $H = f_{xx}(1,1)f_{yy}(1,1) - [f_{xy}(1,1)]^2 = (2)(2) - 0 = 4$. Since H is positive and $f_{xx}(1,1) = 2$ is positive, by the second-derivative test we know that f has a relative minimum at $(1, 1)$. That is, the point $(1, 1, -2)$ is a point in 3-dimensional space that corresponds to a relative minimum point on the surface defined by $z = x^2 - 2x + y^2 - 2y$. To investigate the region about the point $(1,1)$, fill in your worksheet with the values -1 to 3 in cells B1:F1 and -1 to 3 in cells A2:A6. Enter the formula, =B1^2-2*B1+$A2^2-2*$A2, into cell B2 and use it to generate the required values:

	A	B	C	D	E	F
1		-1	0	1	2	3
2	-1	6	3	2	3	6
3	0	3	0	-1	0	3
4	1	2	-1	-2	-1	2
5	2	3	0	-1	0	3
6	3	6	3	2	3	6

Note that as we move away from $(1, 1)$ the function values increase, indicating that we are at a relative minimum. The following plot confirms that the point $(1, 1, -2)$ is a relative minimum.

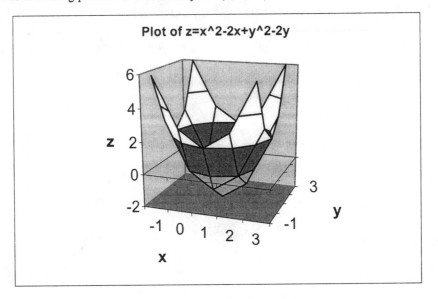

Plot of z=x^2-2x+y^2-2y

8-5 Constrained Maxima and Minima and Applications

11. Find the maximum and minimum values and the points at which they occur for the following function:

$$f(x, y) = x^2 + y^2 \ , \ 0 \leq x \leq 2, \ 0 \leq y \leq 2$$

Verify the solution using Excel's Solver.

Solution: First use the second-derivative test to check for relative maxima and minima. The necessary derivatives are $f_x = 2x, f_{xx} = 2, f_y = 2y, f_{yy} = 2,$ and $f_{xy} = 0$. The point, (0, 0), is a critical point. The quantity, H, is equal to $H = f_{xx}(0,0) f_{yy}(0,0) - [f_{xy}(0,0)]^2 = (2)(2) - 0 = 4$. Since H is positive and $f_{xx}(0,0) = 2$ is positive, by the second-derivative test we know that f has a relative minimum at (0, 0). The function value here is 0. On the boundary, the maximum value of $f(x, y)$ is 8, and it occurs when $x = 2$ and $y = 2$. The minimum function value is 0 and it occurs at (0, 0). To use Excel's Solver, first set up the spreadsheet as follows.

	A	B	C	D
1	1	1	1	0
2			1	2
3			1	0
4			1	2
5			2	

	A	B	C	D
1	1	1	A1	0
2			A1	2
3			B1	0
4			B1	2
5			A1^2+B1^2	

Enter into cells A1 and B1 an initial guess for the x and y solution for the coordinates of the minimum. Then enter the constraints are into cells C1:D4. Now enter the formula, =A1^2+B2^2, into cell C5. From the Tools menu, bring up the Solver Parameters dialog box. In the Solver Parameters dialog box, enter C5 in the Set Target Cell text box and choose the Equal To Min option if it is not already chosen. Enter A1:B1 into the By Changing Cells text box. For the Subject to the Constraints list box, click the Add button, and enter the constraints by following the directions. The completed Solver Parameters dialog box appears as follows:

After clicking the Solve button, the following Solver Results dialog box appears:

In the Reports list box, choose Answer and click the OK button to obtain the following spreadsheet:

	A	B	C	D
1	0	0	0	0
2			0	2
3			0	0
4			0	2
5			0	

The 0 in A1 and the 0 in B1 tell us that the minimum occurs at (0, 0). The 0 in cell C5 tells us that the minimum function value is 0.

The solution for the maximum follows the same procedure, except in the first dialog box select the Equal to Max option instead of Min. The completed Solver Parameters dialog box and Solver's solution appear in the following two panels:

	A	B	C	D
1	2	2	2	0
2			2	2
3			2	0
4			2	2
5			8	

The new spreadsheet now tells us that the maximum occurs at (2, 2) and the maximum value is 8. This confirms the values we found analytically.

8-6 Double Integrals

12. Consider the double integral, $\int_0^1 \int_0^1 5\,dx\,dy$. Find the value of this double integral. Then interpret the integral and approximate its value by using a double Riemann sum.

Solution: $\int_0^1 \int_0^1 5\,dx\,dy = \int_0^1 [5x]_0^1\,dy = \int_0^1 5\,dy = [5y]_0^1 = 5$. The double integral over the region described by $0 \le x \le 1$ and $0 \le y \le 1$ in the xy-plane beneath the plane, $z = 5$, determines the volume of a box. The dimensions of the box are 1 by 1 by 5 and we know, therefore, that the volume is 5.

The double Riemann sum can be applied as follows. Suppose the region of integration is partitioned as follows. Suppose $0 \le x \le 1$ is partitioned into 0, 0.5, and 1 and $0 \le y \le 1$ is partitioned into 0, 0.5, and 1. Then $\Delta x = 0.5$ and $\Delta y = 0.5$. The region of integration is broken into four squares. The first square is $0 \le x \le 0.5$, $0 \le y \le 0.5$. The area of this square is $\Delta x\,\Delta y = 0.5(0.5) = 0.25$. Then $f(0, 0) = 5$ is the height to the surface above this square, and $f(0, 0)\,\Delta x\,\Delta y = 5(0.25) = 1.25$ is the volume under the plane, $z = 5$, that corresponds to this region. This same discussion holds for all four squares. The Riemann sum is

$[f(0, 0)\,\Delta x\,\Delta y + f(0, 0.5)\,\Delta x\,\Delta y + f(0.5, 0)\,\Delta x\Delta y + f(0.5, 0.5)\,\Delta x\,\Delta y]$

$$= [f(0, 0) + f(0, 0.5) + f(0.5, 0) + f(0.5, 0.5)](\Delta x\,\Delta y) = [5 + 5 + 5 + 5](0.25) = 5$$

This corresponds to expressing the area of the 1 by 1 by 5 box as the sum of four 0.5 by 0.5 by 5 boxes. This always happens when the surface is parallel to the xy-plane as is the case with the surface $z = 5$. If the region is broken into 25 squares, each of size 0.2 by 0.2, as shown in the following spreadsheet, then we need to sum all the function values in the sheet and multiply the result by $\Delta x\,\Delta y = 0.2\,(0.2) = 0.04$.

	A	B	C	D	E	F	G
1		0	0.2	0.4	0.6	0.8	
2	0	5	5	5	5	5	
3	0.2	5	5	5	5	5	
4	0.4	5	5	5	5	5	
5	0.6	5	5	5	5	5	
6	0.8	5	5	5	5	5	
7		SUM(B2:B6)	SUM(C2:C6)	SUM(D2:D6)	SUM(E2:E6)	SUM(F2:F6)	SUM(B7:F7)

The result of the sums appears in the next panel:

	A	B	C	D	E	F	G
1		0	0.2	0.4	0.6	0.8	
2	0	5	5	5	5	5	
3	0.2	5	5	5	5	5	
4	0.4	5	5	5	5	5	
5	0.6	5	5	5	5	5	
6	0.8	5	5	5	5	5	
7		25	25	25	25	25	125

The Riemann sum in this case is $125(0.04) = 5$. This corresponds to expressing the area of the 1 by 1 by 5 box as the sum of 25 boxes with dimensions 0.2 by 0.2 by 5.

13. Consider the double integral $\int_0^2 \int_0^2 (5x - 2y + 4)\ dxdy$. Find the value of this double integral. Then interpret the integral and approximate its value by using a double Riemann sum. Parallel the technique illustrated in problem 12 of this section to find the Riemann sum.

Solution: $\int_0^2 \int_0^2 (5x - 2y + 4)\ dxdy = \int_0^2 \left[2.5x^2 - 2xy + 4x\right]_0^2 dy = \int_0^2 18 - 4y\ dy = \left[18y - 2y^2\right]_0^2 = 28$. The region, R, described by $0 \leq x \leq 2$, $0 \leq y \leq 2$ in the xy-plane beneath the plane, $z = 5x - 2y + 4$, determines a volume of 28 cubic units. Enter the partition points 0, 0.4, 0.8, 1.2, and 1.6 for x and y into the spreadsheet using row 1 and column A. Then enter the formula, =5*B$1-2*$A2+4, into cell B2 and use it to obtain the required function values. Enter the formula, =SUM(B2:F6), and Excel returns the grand sum, 160, to cell G7:

	A	B	C	D	E	F	G
1		0	0.4	0.8	1.2	1.6	
2	0	4	6	8	10	12	
3	0.4	3.2	5.2	7.2	9.2	11.2	
4	0.8	2.4	4.4	6.4	8.4	10.4	
5	1.2	1.6	3.6	5.6	7.6	9.6	
6	1.6	0.8	2.8	4.8	6.8	8.8	
7		12	22	32	42	52	160

Therefore, the double Riemann sum approximation to the integral is $160(0.4)(0.4) = 25.6$. Now suppose we wish to refine the double Riemann sum by creating more partitions. Suppose we take $\Delta x = \Delta y = 0.05$. Enter 0 in cell B1 and select this cell. Now use the pull-down sequence, Edit → Fill → Series to invoke the Series dialog box. In this dialog box, choose Series in Rows, Step value 0.05, Type as Linear, and Stop value as 1.95.

Excel fills in the numbers 0 through 1.95 in cells B1:AO. A similar operation for Series in Columns fills in the numbers 0 through 1.95 in cells A2:A41. Now enter the formula, =5*B$1-2*$A2+4, into cell B2 and use it to complete the spreadsheet. The grand sum is found to equal 11,119. The Riemann sum approximation to the integral is $11{,}119\Delta x\,(\Delta y) = 11{,}119(0.05)(0.05) = 27.80$. Note how much closer to the true value of 28 we are than when we used $\Delta x = \Delta y = 0.4$. If the refinement $\Delta x = \Delta y = 0.01$ is used, the grand sum is found to be 279,599 and the double Riemann sum approximation is $279{,}599\Delta x(\Delta y) = 279{,}599(0.01)(0.01) = 27.9599$, which is even closer to the true value, 28. We see that the double Riemann sums converge to the true volume.

14. The double integral, $\displaystyle\int_{-2}^{2}\int_{-2}^{2}\frac{1}{2\pi}e^{-\frac{(x^2+y^2)}{2}}\,dxdy$, is a double integral that cannot be evaluated exactly since the antiderivative of the integrand does not exist. However we can use a double Riemann sum to approximate this double integral. Use the techniques illustrated in problems 1 and 2 of this chapter with $\Delta x = \Delta y = 0.1$ and with $\Delta x = \Delta y = 0.05$.

Solution: First, enter –2 into cell B1 and select the cell. Then use the pull-down sequence Edit → Fill → Series to fill in the data. Complete the Series dialog box to fill in the rest of the row as follows:

Next, enter –2 into cell A2 and use a similar procedure to fill in the rest of the column:

Then enter the formula, =1/(2*PI())*EXP(-(B1^2+A2^2)/2), into cell B2 and obtain the required values in cells B2:AO42. The grand sum is 91.0726, found by the formula, =SUM(B2:AO42). The double Riemann sum approximation is $91.0726\Delta x(\Delta y) = 91.0726(0.1)(0.1) \approx 0.9107$.

Using a similar procedure, complete the Series dialog boxes required to obtain the row and column for $\Delta x = \Delta y = 0.05$ as follows:

Using the same formula, obtain the required values in cells B2:C81. Finally, compute the grand sum, 364.3935, using the formula, =SUM(B2:CC81). Therefore, the double Riemann sum approximation is $364.3935\Delta x(\Delta y) = 364.3935 (0.05)(0.05) \approx 0.9110$.

Chapter Review Exercises

1. Plot the interaction function, $z = x + y + xz - 3$, for values of x between 0 and 2 and values of y between 0 and 3.

2. Plot the circle centered at the origin and having radius 1.

3. A sphere described by the equation, $x^2 + y^2 + (z - 5)^2 = 25$, has a radius of $r = 5$ and is centered on the z-axis at $(0, 0, 5)$. Construct that portion of the top half of the surface of the sphere above a square in the xy-plane with corners at $(1, 1)$, $(-1, 1)$, $(-1, -1)$, and $(1, -1)$.

4. Use the second-derivative test to locate the relative maxima and minima of the function defined by $z = \sqrt{2y - x^2 - y^2}$. Then use Excel to investigate the neighborhood surrounding the points.

5. Find the value of the double integral, $\displaystyle\int_0^2\int_0^2 x^2 + y^2 \; dxdy$. Then give an interpretation for this double integral and approximate its value by using a double Riemann sum.

6. The double integral, $\displaystyle\int_0^2\int_0^2 e^{-x^2-y^3} \; dxdy$, cannot be evaluated exactly since the antiderivative of the integrand does not exist. Use a double Riemann sum to approximate this double integral with $\Delta x = \Delta y = 0.1$ and with $\Delta x = \Delta y = 0.05$.

APPENDIX

Solutions to Chapter Review Questions

Linear Functions and Models

1.1 The required spreadsheet with the values in target cells B1:B10 returned by the formula, =3*A1^2+5*A1-17:

	A	B
1	6	121
2	7	165
3	8	215
4	9	271
5	10	333
6	11	401
7	12	475
8	13	555
9	14	641
10	15	733

1.2 First write the function that describes ABC's compensation plan. Let $C(s)$ = the commission amount paid as a function of the sales volume, s.

$$C(s) = \begin{cases} 0.05s & \text{if } 0 \leq s \leq 1,000,000 \\ 50,000 + 0.06(s - 1,000,000) & \text{if } 1,000,000 < s \leq 5,000,000 \\ 290,000 + 0.07(s - 5,000,000) & \text{if } s > 5,000,000 \end{cases}$$

Cells A1:B3 should be formatted as currency, and the equivalent Excel formula is:

=IF(A1>5000000,290000+0.07*(A1-5000000), IF(A1>1000000,50000+0.06*(A1-1000000),0.05*A1)).

When the above formula is executed using the data in cells A1:A3, Excel returns to cells B1:B3:

	A	B	C
1	$750,000	$37,500	Alice
2	$3,600,000	$206,000	Bob
3	$9,200,000	$584,000	Charles
4	Sales Volume	Commission	

1.3 The required spreadsheet data:

	A	B
1	$200,000	$10,000
2	$400,000	$20,000
3	$600,000	$30,000

	A	B
14	$10,000,000	$640,000
15	$11,500,000	$745,000

The graph is constructed using the Chart Wizard:

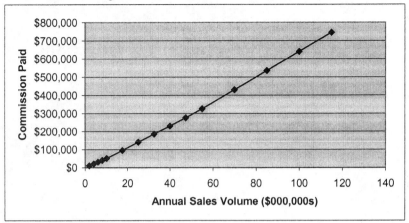

1.4 The required spreadsheet:

	A	B	C
1	-2	0	2.5
2	0	5	2.5
3	2	10	2.5
4	4	15	2.5
5	6	20	2.5
6	8	25	2.5
7	10	30	2.5
8	12	35	2.5
9	14	40	2.5
10	16	45	
11	x	y	slope

The function is indeed linear. A constant slope of 2.5 exists between every successive pair of values of x and y. The y-intercept value is 5, which is the value of y when $x = 0$. Therefore, the equation for the line is $y = 2.5x + 5$.

1.5 The formula for Alex's total cost, revenue, and profit:

$C = 300 + 20Q$

$R = 50Q$

$P = R - C = 50Q - (300 + 20Q) = 30Q - 300$

The required spreadsheet:

	A	B	C	D
1	0	$300	$0	-$300
2	10	$500	$500	$0
3	20	$700	$1,000	$300
4	30	$900	$1,500	$600
5	40	$1,100	$2,000	$900
6	Units	Cost	Revenue	Profit

126

Alex will break even if he sells 10 drumsticks each month. At break even, both his total cost and revenue equal $500. If he could build and sell 30 drumsticks per month, his monthly profit would be $600. The required chart using the Chart Wizard:

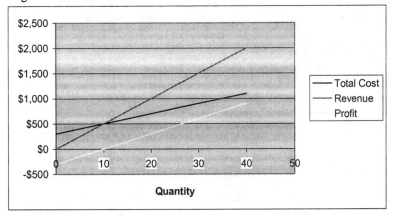

1.6 The spreadsheet with the LINEST( ) array function output returned to cells D1:E5:

	A	B	C	D	E
1	26.1	2500		-0.006117	40.41312
2	24.2	2550		0.00046	1.565331
3	22.4	2700		0.94653	1.13751
4	18.3	3500		177.0195	10
5	13.4	4300		229.0507	12.93929
6	19.5	3100			
7	11.0	4800			
8	20.1	3400			
9	25.6	2600			
10	23.7	2900			
11	19.1	3600			
12	17.2	4000			

a) Slope: –0.006117; Intercept: 40.41312
b) At x = 3900 lb, Predicted fuel economy = (–0.006117)(3900) + 40.41312 = 16.56 mpg.
c) Correlation coefficient: 0.9729
d) The required graph and scatter plot:

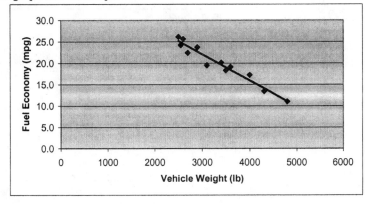

Non-Linear Functions and Models

2.1 The intercepts are $x = 2$ and 2. The curve is tangent to the x-axis at $x = 2$. The curve is concave downward. The vertex is located at $(2, 0)$ and the line of symmetry is the line $x = 2$.

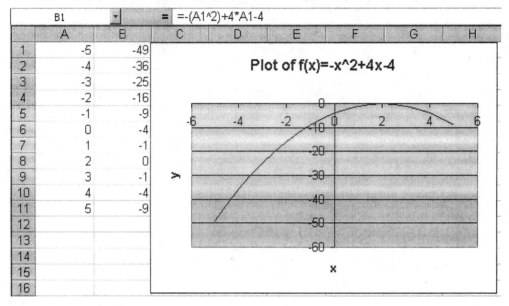

2.2 As the number of advertisements increases, the percentage of subscribers approaches 65%.

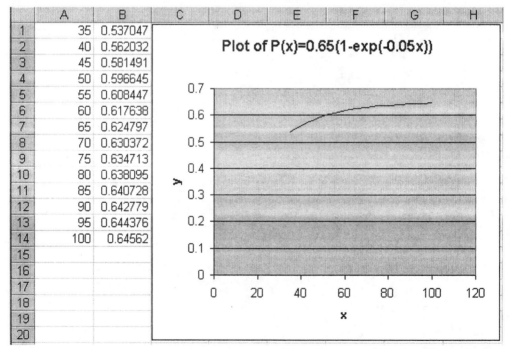

2.3

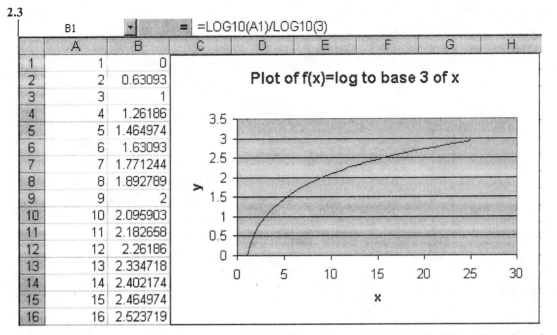

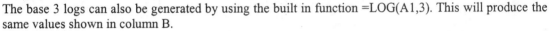

The base 3 logs can also be generated by using the built in function =LOG(A1,3). This will produce the same values shown in column B.

2.4

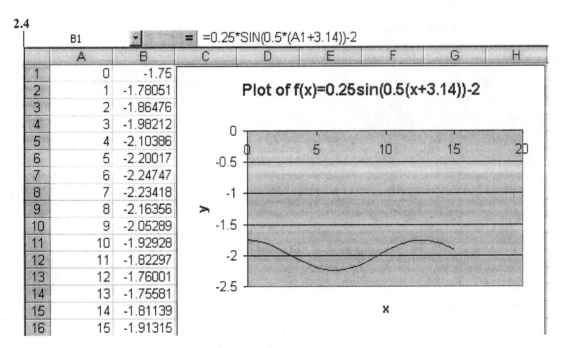

2.5 The required spreadsheet and the chart:

	B1		▼		= =COS(A1)/SIN(A1)

	A	B	C	D	E	F	G
1	0.1	9.966644					
2	0.2	4.933155			**Plot of f(x) = cotx**		
3	0.3	3.232728					
4	0.4	2.365222					
5	0.5	1.830488					
6	0.6	1.461696					
7	0.7	1.187242					
8	0.8	0.971215					
9	0.9	0.793551					
10	1	0.642093					
11	1.1	0.508968					
12	1.2	0.38878					
13	1.3	0.277616					
14	1.4	0.172477					
15	1.5	0.070915					
16	1.6	-0.02921					
17	1.7	-0.12993					

Introduction to the Derivative

3.1 The required spreadsheet:

	A	B	C	D	E
1	year	CD's shipped	average rate of change		average rate of change
2	1995	723	with respect to 1995		from year to year
3	1996	779	56		56
4	1997	759	18		-20
5	1998	847	41.33333333		88
6	1999	939	54		92

3.2 The required spreadsheet and the chart:

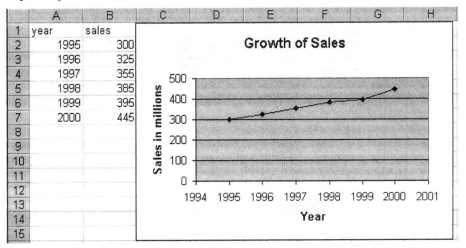

3.3 The average rate of change is approaching 8 as *h* approaches 0.

A2		=	=(50+2*(2+A1)^2-58)/A1	
	A	B	C	D	E
1	1	0.1	0.01	0.001	0.0001
2	10	8.2	8.02	8.002	8.0002

3.4 The derivative of the common log when *x* = 1 is 0.434294.

A2		=	=(LOG10(1+A1)-LOG10(1-A1))/(2*A1)		
	A	B	C	D	E	F
1	0.1	0.01	0.001	0.0001	0.00001	
2	0.435751	0.434309	0.434295	0.434294	0.434294	

3.5 The following is a plot of $f(x) = (x-1)^3$.

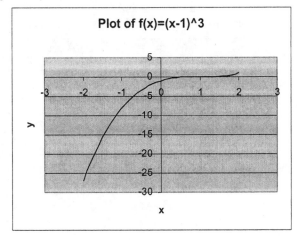

The slopes of the secant lines approach the slope of the tangent line at $(0, -1)$. The slopes of the secant lines are approaching 3. The slope of the tangent line at $(0, -1)$ is 3 since the derivative, $3(x-1)^2$, when evaluated at $x = 0$, is 3.

A2			=	=((A1-1)^3-(-A1-1)^3)/(2*A1)	
	A	B	C	D	E
1	1	0.1	0.01	0.001	0.0001
2	4	3.01	3.0001	3.000001	3

3.6 The derivative is $\dfrac{1}{2\sqrt{x}}$. Note that the function is undefined for negative values of x and the derivative is undefined for $x \leq 0$.

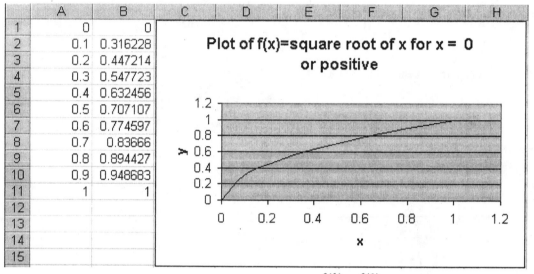

As h approaches 0 from the positive direction, the ratio $\dfrac{f(h) - f(0)}{h}$ grows larger and larger. The ratio as h approaches 0 from the right is undefined, since $f(h)$ is undefined for $h < 0$. The tangent line becomes vertical as x approaches 0 from the right and the slope grows larger and larger.

A2			=	=SQRT(A1)/A1	
	A	B	C	D	E
1	1	0.1	0.01	0.001	0.0001
2	1	3.162278	10	31.62278	100

3.7 The derivative of $P(x)$ is $-\dfrac{x}{10,000} + 2.44$. The derivative is 0 when $x = 24,400$. The maximum value of $P(x)$ occurs when $x = 24,400$ and this maximum value is 24,768.

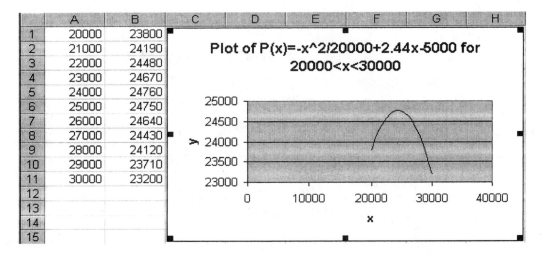

Plot of P(x)=-x^2/20000+2.44x-5000 for 20000<x<30000

	A	B
1	20000	23800
2	21000	24190
3	22000	24480
4	23000	24670
5	24000	24760
6	25000	24750
7	26000	24640
8	27000	24430
9	28000	24120
10	29000	23710
11	30000	23200
12		
13		
14		
15		

3.8 As can be seen from the following spreadsheet, the limit is e.

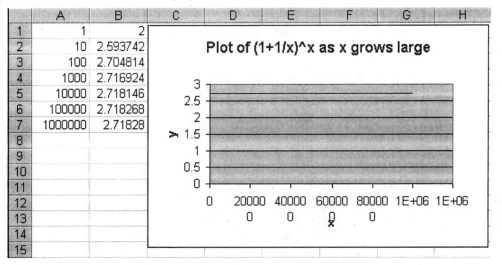

Plot of (1+1/x)^x as x grows large

	A	B
1	1	2
2	10	2.593742
3	100	2.704814
4	1000	2.716924
5	10000	2.718146
6	100000	2.718268
7	1000000	2.71828
8		
9		
10		
11		
12		
13		
14		
15		

3.9 The following spreadsheet indicates that the one-sided limits are different at $x = 0$ and $x = 5$. This function is discontinuous at both of these points.

	A	B	C	D
1	x	f(x)	x	f(x)
2	-1	1	4	4
3	-0.1	1	4.9	4.9
4	-0.01	1	4.99	4.99
5	-0.001	1	4.999	4.999
6	-0.0001	1	4.9999	4.9999
7	0.0001	0.0001	5.0001	2.23609
8	0.001	0.001	5.001	2.236292
9	0.01	0.01	5.01	2.238303
10	0.1	0.1	5.1	2.258318
11	1	1	6	2.44949

Techniques of Differentiation

4.1 The derivative is $7x^6 - 6x^5 + 4x^3 - 3x^2$. This derivative evaluated at $x = 2$ is 2.76·

	A	B	C	D
1	h	f(2-h)	f(2+h)	(f(2+h)-f(2-h))/2h
2	0.1	((2-A2)^3-(2-A2)^2)*((2-A2)+(2-A2)^4)	((2+A2)^3-(2+A2)^2)*((2+A2)+(2+A2)^4)	(C2-B2)/(2*A2)
3	0.01	((2-A3)^3-(2-A3)^2)*((2-A3)+(2-A3)^4)	((2+A3)^3-(2+A3)^2)*((2+A3)+(2+A3)^4)	(C3-B3)/(2*A3)
4	0.001	((2-A4)^3-(2-A4)^2)*((2-A4)+(2-A4)^4)	((2+A4)^3-(2+A4)^2)*((2+A4)+(2+A4)^4)	(C4-B4)/(2*A4)
5	0.0001	((2-A5)^3-(2-A5)^2)*((2-A5)+(2-A5)^4)	((2+A5)^3-(2+A5)^2)*((2+A5)+(2+A5)^4)	(C5-B5)/(2*A5)

	A	B	C	D
1	h	f(2-h)	f(2+h)	(f(2+h)-f(2-h))/2h
2	0.1	48.51439	104.5298	280.077201
3	0.01	69.2846	74.80541	276.0407007
4	0.001	71.72445	72.27645	276.000407
5	0.0001	71.9724	72.0276	276.0000041

4.2 The derivative is $\dfrac{-2x^3 + 4}{(x^3 + 4)^2}$ and its value at $x = 1$ is 0.08.

	A	B	C	D
1	h	f(1-h)	f(1+h)	(f(1+h)-f(1-h))/2h
2	0.1	(1-A2)/((1-A2)^3+4)	(1+A2)/((1+A2)^3+4)	(C2-B2)/(2*A2)
3	0.01	(1-A3)/((1-A3)^3+4)	(1+A3)/((1+A3)^3+4)	(C3-B3)/(2*A3)
4	0.001	(1-A4)/((1-A4)^3+4)	(1+A4)/((1+A4)^3+4)	(C4-B4)/(2*A4)
5	0.0001	(1-A5)/((1-A5)^3+4)	(1+A5)/((1+A5)^3+4)	(C5-B5)/(2*A5)

	A	B	C	D
1	h	f(1-h)	f(1+h)	(f(1+h)-f(1-h))/2h
2	0.1	0.190315	0.20634	0.080125983
3	0.01	0.199183	0.200783	0.08000128
4	0.001	0.19992	0.20008	0.080000013
5	0.0001	0.199992	0.200008	0.08

4.3 The derivative is $4(1+x)^3$ and its value at $x = 2$ is 108.

	A	B	C	D
1	h	f(2-h)	f(2+h)	(f(2+h)-f(2-h))/2h
2	0.1	(1+(2-A2))^4	(1+(2+A2))^4	(C2-B2)/(2*A2)
3	0.01	(1+(2-A3))^4	(1+(2+A3))^4	(C3-B3)/(2*A3)
4	0.001	(1+(2-A4))^4	(1+(2+A4))^4	(C4-B4)/(2*A4)
5	0.0001	(1+(2-A5))^4	(1+(2+A5))^4	(C5-B5)/(2*A5)
6				

1	h	f(2-h)	f(2+h)	(f(2+h)-f(2-h))/2h
2	0.1	70.7281	92.3521	108.12
3	0.01	79.92539	82.08541	108.0012
4	0.001	80.89205	81.10805	108.000012
5	0.0001	80.9892	81.0108	108.0000001

4.4 The derivative is $\dfrac{2x}{x^2+4}$ and its value at $x = 3$ is 0.4615.

	A	B	C	D
1	h	f(3-h)	f(3+h)	(f(3+h)-f(3-h))/2h
2	0.1	LN((3-A2)^2+4)	LN((3+A2)^2+4)	(C2-B2)/(2*A2)
3	0.01	LN((3-A3)^2+4)	LN((3+A3)^2+4)	(C3-B3)/(2*A3)
4	0.001	LN((3-A4)^2+4)	LN((3+A4)^2+4)	(C4-B4)/(2*A4)
5	0.0001	LN((3-A5)^2+4)	LN((3+A5)^2+4)	(C5-B5)/(2*A5)

	A	B	C	D
1	h	f(3-h)	f(3+h)	(f(3+h)-f(3-h))/2h
2	0.1	2.518503	2.610805	0.461511087
3	0.01	2.560331	2.569562	0.461538188
4	0.001	2.564488	2.565411	0.461538459
5	0.0001	2.564903	2.564996	0.461538462

4.5

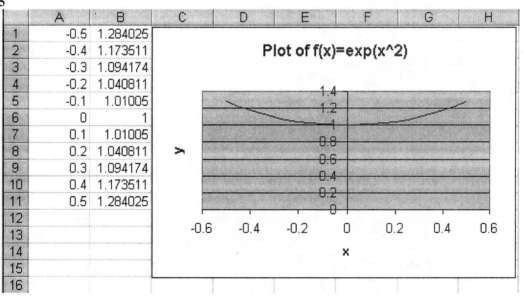

The derivative is $2xe^{x^2}$ and its value at $x = 0$ is 0. The derivative using the limit definition is as follows:

	A	B	C	D
1	h	f(0-h)	f(0+h)	(f(0+h)-f(0-h))/2h
2	0.1	EXP(-(A2^2))	EXP(A2^2)	(C2-B2)/(2*A2)
3	0.01	EXP(-(A3^2))	EXP(A3^2)	(C3-B3)/(2*A3)
4	0.001	EXP(-(A4^2))	EXP(A4^2)	(C4-B4)/(2*A4)
5	0.0001	EXP(-(A5^2))	EXP(A5^2)	(C5-B5)/(2*A5)
6	0.00001	EXP(-(A6^2))	EXP(A6^2)	(C6-B6)/(2*A6)

	A	B	C	D
1	h	f(0-h)	f(0+h)	(f(0+h)-f(0-h))/2h
2	0.1	0.990049834	1.010050167	0.100001667
3	0.01	0.999900005	1.000100005	0.01
4	0.001	0.999999	1.000001	0.001
5	0.0001	0.99999999	1.00000001	1E-04
6	0.00001	1	1	1E-05

4.6

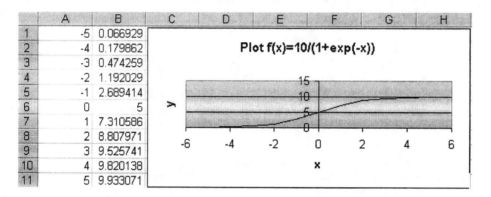

	A	B
1	-5	0.066929
2	-4	0.179862
3	-3	0.474259
4	-2	1.192029
5	-1	2.689414
6	0	5
7	1	7.310586
8	2	8.807971
9	3	9.525741
10	4	9.820138
11	5	9.933071

The derivative is $\dfrac{10e^{-x}}{(1+e^{-x})^2}$ and its value at $x = 10$ is 0.00000454. This tells us that the curve is almost horizontal at $x = 10$. The limit of $f(x)$ as x approaches infinity is 10.

4.7

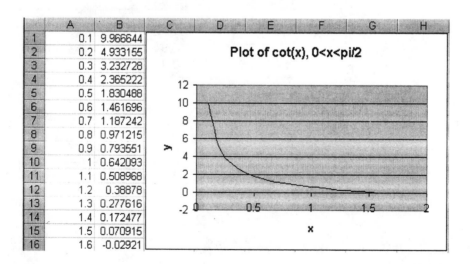

	A	B
1	0.1	9.966644
2	0.2	4.933155
3	0.3	3.232728
4	0.4	2.365222
5	0.5	1.830488
6	0.6	1.461696
7	0.7	1.187242
8	0.8	0.971215
9	0.9	0.793551
10	1	0.642093
11	1.1	0.508968
12	1.2	0.38878
13	1.3	0.277616
14	1.4	0.172477
15	1.5	0.070915
16	1.6	-0.02921

The derivative of the cotangent function is $\dfrac{-1}{(\sin x)^2}$. A plot of this derivative is as follows:

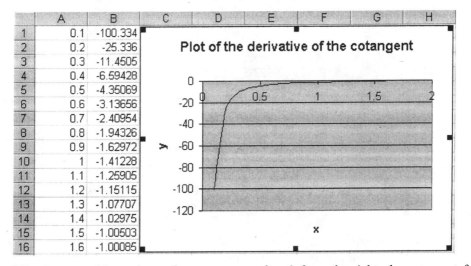

	A	B
1	0.1	-100.334
2	0.2	-25.336
3	0.3	-11.4505
4	0.4	-6.59428
5	0.5	-4.35069
6	0.6	-3.13656
7	0.7	-2.40954
8	0.8	-1.94326
9	0.9	-1.62972
10	1	-1.41228
11	1.1	-1.25905
12	1.2	-1.15115
13	1.3	-1.07707
14	1.4	-1.02975
15	1.5	-1.00503
16	1.6	-1.00085

Plot of the derivative of the cotangent

The following spreadsheet shows that as x approaches 0 from the right, the cotangent function approaches plus infinity:

B1   = =COS(A1)/SIN(A1)

	A	B	C	D	E
1	0.1	9.966644			
2	0.01	99.99667			
3	0.001	999.9997			
4	0.0001	10000			
5	0.00001	100000			

4.8

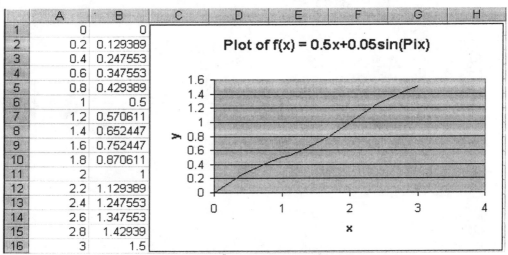

	A	B
1	0	0
2	0.2	0.129389
3	0.4	0.247553
4	0.6	0.347553
5	0.8	0.429389
6	1	0.5
7	1.2	0.570611
8	1.4	0.652447
9	1.6	0.752447
10	1.8	0.870611
11	2	1
12	2.2	1.129389
13	2.4	1.247553
14	2.6	1.347553
15	2.8	1.42939
16	3	1.5

Plot of f(x) = 0.5x+0.05sin(Pix)

The derivative of the function is $0.5 + 0.05\pi\cos(\pi x)$ and its value at $x = 1.5$ is 0.5.

	A	B	C	D
1	h	f(1.5-h)	f(1.5+h)	(f(1.5+h)-f(1.5-h))/2h
2	0.1	0.652447	0.752447	0.499999385
3	0.01	0.695025	0.705025	0.499999375
4	0.001	0.6995	0.7005	0.499999375
5	0.0001	0.69995	0.70005	0.499999375

The same value is obtained for the derivative at $x = 1.5$ as when using the rules of differentiation.

4.9 Using implicit differentiation, we have $2x + 8y\dfrac{dy}{dx} = 0$ or $\dfrac{dy}{dx} = -\dfrac{x}{4y}$. The slope of the tangent line

at $(\sqrt{2}, -\dfrac{\sqrt{2}}{2})$ is $-\dfrac{\sqrt{2}}{4\left(-\dfrac{\sqrt{2}}{2}\right)} = 0.5$. The equation of the lower half of the ellipse is $y = -\sqrt{1 - \left(\dfrac{x}{2}\right)^2}$.

Using this equation and the limit definition of the derivative, we obtain the same value for the derivative at $x = 1.414$, the approximate square root of 2.

	A	B	C	D
1	h	f(1.414-h)	f(1.414+h)	(f(1.414+h)-f(1.414-h))/2h
2	0.1	-0.75389	-0.653414876	0.502378499
3	0.01	-0.71218	-0.702179464	0.499874003
4	0.001	-0.70771	-0.706713344	0.499849273
5	0.0001	-0.70726	-0.707163558	0.499849025

Applications of the Derivative

5.1

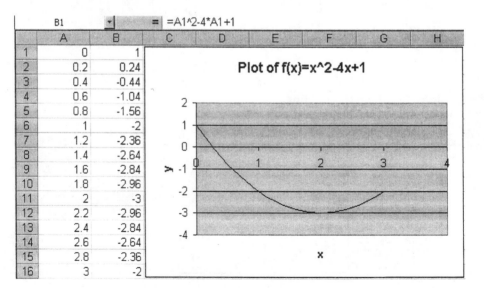

The point $(0, 1)$ corresponds to the absolute maximum, the point $(2, -3)$ corresponds to the absolute minimum, and the point $(3, -2)$ corresponds to a relative maximum on the domain $[0, 3]$.

5.2 The graph of $f(x) = x^3 + x$ is as follows:

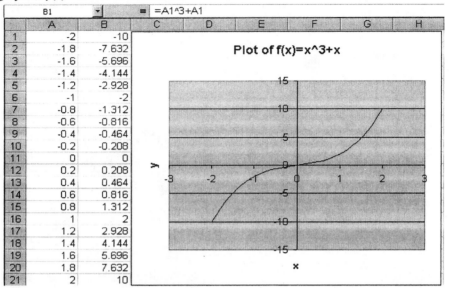

From the graph, it appears that the absolute minimum is located at $(-2, -10)$ and the absolute maximum occurs at $(2, 10)$. The derivative of $f(x)$ is $3x^2 + 1$. The derivative is never 0, so there are no critical points.

5.3 The area is $A = LW$. The constraint $2(4L) + 2(2W) = 80$ exists because of the budget constraint. Solving the constraint equation for L, we get $L = 10 - 0.5W$. Substituting in the area equation we have $A = (10 - 0.5W)W = 10W - 0.5W^2$. The following is a plot of the area function versus different widths:

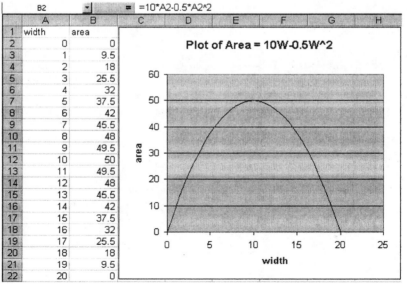

The graph reveals that the largest area measures 50 square units. The calculus approach is to take the derivative of the area with respect to the width. The result is $10 - W$. Setting this derivative equal to 0, we find $W = 10$. From the equation $L = 10 - 0.5W$, we find $L = 5$. The maximum area occurs when $W = 10$ and $L = 5$ and is equal to $A = 5(10) = 50$, the same result we found graphically.

5.4 A plot of the function is as follows:

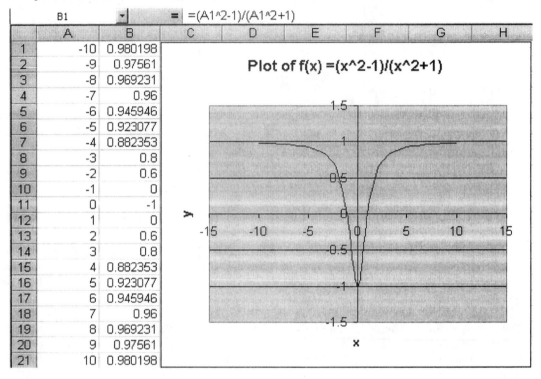

	A	B	C	D	E	F	G	H
B1		▼		=	=(A1^2-1)/(A1^2+1)			
1	-10	0.980198						
2	-9	0.97561						
3	-8	0.969231						
4	-7	0.96						
5	-6	0.945946						
6	-5	0.923077						
7	-4	0.882353						
8	-3	0.8						
9	-2	0.6						
10	-1	0						
11	0	-1						
12	1	0						
13	2	0.6						
14	3	0.8						
15	4	0.882353						
16	5	0.923077						
17	6	0.945946						
18	7	0.96						
19	8	0.969231						
20	9	0.97561						
21	10	0.980198						

Plot of f(x) =(x^2-1)/(x^2+1)

The x-intercepts are $(-1, 0)$ and $(1, 0)$. The y-intercept is $(0, -1)$. The first derivative is $\dfrac{4x}{(x^2+1)^2}$ and the second derivative is $-4\dfrac{3x^2-1}{(x^2+1)^3}$. The first derivative is 0 when $x = 0$. The second derivative is positive when $x = 0$. Therefore $x = 0$ corresponds to a minimum. The function approaches 1 as x approaches minus infinity and as x approaches plus infinity.

5.5 The area of a circle is $A = \pi R^2$. The rate of change of the area is related to the rate of change of the radius by the equation $\dfrac{dA}{dt} = 2\pi R \dfrac{dR}{dt}$. The table relating the two rates of change is as follows:

	B1			=	=31.41593*A1

	A	B	C	D
1	1	31.41593	6	188.4956
2	1.5	47.1239	6.5	204.2035
3	2	62.83186	7	219.9115
4	2.5	78.53983	7.5	235.6195
5	3	94.24779	8	251.3274
6	3.5	109.9558	8.5	267.0354
7	4	125.6637	9	282.7434
8	4.5	141.3717	9.5	298.4513
9	5	157.0797	10	314.1593
10	5.5	172.7876		

5.6 The revenue function $R = pq = p(1000 - 10p) = 1000p - 10p^2$. The elasticity of demand is defined as

$$E = -\frac{dq}{dp} \cdot \frac{p}{q} = -(-10)\frac{p}{1000 - 10p} = \frac{p}{100 - p} \, .$$

	A	B	C	D
1	p	q	R	E
2	0	1000	0	0
3	10	900	9000	0.111111
4	20	800	16000	0.25
5	30	700	21000	0.428571
6	40	600	24000	0.666667
7	50	500	25000	1
8	60	400	24000	1.5
9	70	300	21000	2.333333
10	80	200	16000	4
11	90	100	9000	9

The three plots are as follows:

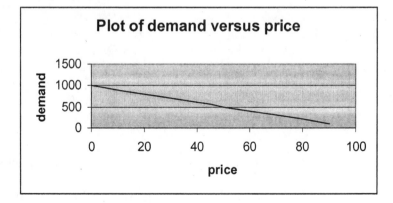

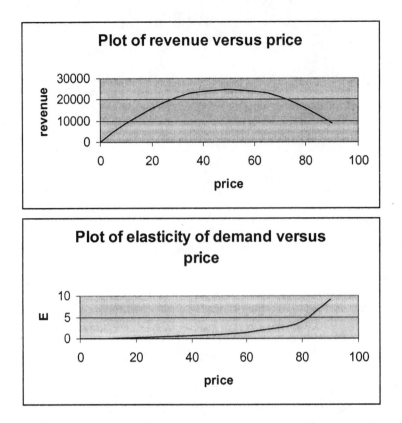

The Integral

6.1 $F(x) = e^x + 5\ln x - \dfrac{4}{3}x^3$ if we select the constant of integration equal to zero. The value of $f(x)$ at

$x = 2$ is $f(2) = e^2 + 5/2 - 4(4) = -6.11094$.

	C2		= =EXP(2-A2)+5*LN(2-A2)-1.33333*(2-A2)^3		
	A	B	C	D	E
1	h	F(2+h)	F(2-h)	(F(2+h)-F(2-h))/2h	
2	1	-10.4213	1.384951828	-5.903131731	
3	0.1	-0.47211	0.749853403	-6.109829485	
4	0.01	0.12655	0.248767891	-6.110893249	
5	0.001	0.182036	0.194257975	-6.110903795	
6	0.0001	0.187541	0.188763043	-6.1109039	

6.2 $F(x) = \dfrac{(3x^2 + 3)^4}{24}$ if we select the constant of integration equal to zero. The value of $f(x)$ at $x = 1$ is

$f(1) = (6)^3 = 216$.

| B2 | | ▼ | = | =(3*(1+A2)^2+3)^4/24 |

	A	B	C	D
1	h	F(1+h)	F(1-h)	(F(1+h)-F(1-h))/2h
2	1	2109.375	3.375	1053
3	0.1	80.50871	36.22330533	221.427027
4	0.01	56.20374	51.88266456	216.0540027
5	0.001	54.21643	53.78443146	216.00054
6	0.0001	54.0216	53.97840432	216.0000054

6.3 Using the fundamental theorem of calculus, we find the value of the integral to be $0.5\ln(2) = 0.3466$.

	C1		▼	=	=SUM(B1:B10)*0.1

	A	B	C	D	E
1	0	0	0.320739		
2	0.1	0.09901			
3	0.2	0.192308			
4	0.3	0.275229			
5	0.4	0.344828			
6	0.5	0.4			
7	0.6	0.441176			
8	0.7	0.469799			
9	0.8	0.487805			
10	0.9	0.497238			

	C1		▼	=	=SUM(B1:B100)*0.01

	A	B	C	D	E
1	0	0	0.344065		
2	0.01	0.009999			
3	0.02	0.019992			
4	0.03	0.029973			
5	0.04	0.039936			
6	0.05	0.049875			
7	0.06	0.059785			
8	0.07	0.069659			
9	0.08	0.079491			
10	0.09	0.089277			

	C1		▼	=	=SUM(B1:B1000)*0.001

	A	B	C	D	E
1	0	0	0.346324		
2	0.001	0.001			
3	0.002	0.002			
4	0.003	0.003			
5	0.004	0.004			
6	0.005	0.005			
7	0.006	0.006			
8	0.007	0.007			
9	0.008	0.007999			
10	0.009	0.008999			

6.4 The approximation with $n = 10$ subdivisions is 41.9%.

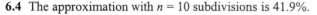

	A	B	C	D	E
1	60	5.33386E-05	0.418681		
2	61	0.000243876			
3	62	0.000950188			
4	63	0.003154734			
5	64	0.008925427			
6	65	0.02151831			
7	66	0.044207931			
8	67	0.077393609			
9	68	0.115457697			
10	69	0.146775499			

B1 = =0.159*EXP(-((A1-70)^2/12.5))

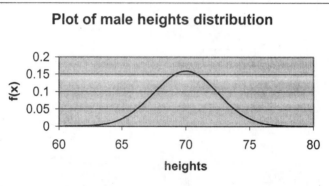

6.5 The integral $\int_0^{2\pi} \cos x \, dx$ is equal to 0, not the area under the curve. To find the area under the curve, you can find the integral with limits from 0 to $\pi/2$ and the integral with limits from $3\pi/2$ to 2π and add these to the negative of the integral with limits from $\pi/2$ to $3\pi/2$. The area is equal to 4 square units. The following is a plot of the cosine curve from $x = 0$ to $x = 2\pi$.

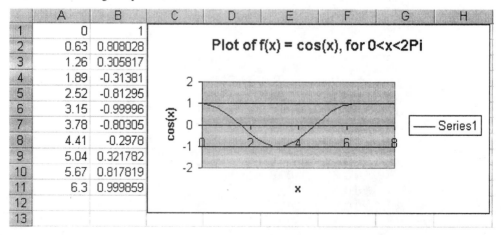

	A	B
1	0	1
2	0.63	0.808028
3	1.26	0.305817
4	1.89	-0.31381
5	2.52	-0.81295
6	3.15	-0.99996
7	3.78	-0.80305
8	4.41	-0.2978
9	5.04	0.321782
10	5.67	0.817819
11	6.3	0.999859
12		
13		

Further Integration Techniques and Applications of the Integral

7.1 The value of $\displaystyle\int_0^1 (x+1)e^x\,dx$, found using integration by parts is $e = 2.718281828$. In the following output, the left Riemann sum is shown in cell C1, the right Riemann sum is shown in cell F1, and the midpoint Riemann sum is shown in cell I1. The midpoint Riemann sum is closest to the correct answer (as is the usual case).

B1   = =(A1+1)*EXP(A1)

	A	B	C	D	E	F	G	H	I
1	0	1	2.501581	0.1	1.215688	2.945238	0.05	1.103835	2.715718
2	0.1	1.215688		0.2	1.465683		0.15	1.336109	
3	0.2	1.465683		0.3	1.754816		0.25	1.605032	
4	0.3	1.754816		0.4	2.088555		0.35	1.915741	
5	0.4	2.088555		0.5	2.473082		0.45	2.274053	
6	0.5	2.473082		0.6	2.91539		0.55	2.686542	
7	0.6	2.91539		0.7	3.42338		0.65	3.160642	
8	0.7	3.42338		0.8	4.005974		0.75	3.70475	
9	0.8	4.005974		0.9	4.673246		0.85	4.328347	
10	0.9	4.673246		1	5.436564		0.95	5.042134	

7.2

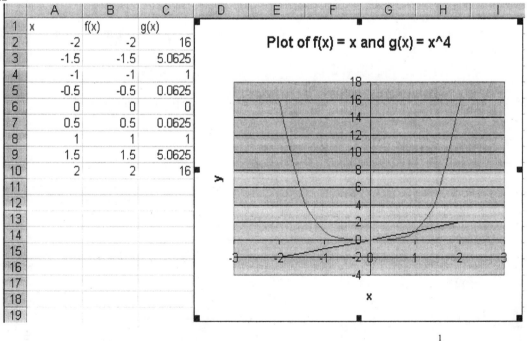

	A	B	C
1	x	f(x)	g(x)
2	-2	-2	16
3	-1.5	-1.5	5.0625
4	-1	-1	1
5	-0.5	-0.5	0.0625
6	0	0	0
7	0.5	0.5	0.0625
8	1	1	1
9	1.5	1.5	5.0625
10	2	2	16

Plot of f(x) = x and g(x) = x^4

The area enclosed by the two graphs extends from $x = 0$ to $x = 1$. The area equals $\displaystyle\int_0^1 (x - x^4)\,dx = 0.3$.

7.3 The average value over $-1 < x < 1$ is $y = \bar{f} = \dfrac{1}{2}\displaystyle\int_{-1}^{1} e^x \, dx = 1.175$.

	A	B	C	D	E	F	G	H	I
1	x	f(x) =exp(x)	y = 1.175						
2	-1	0.36787944	1.175						
3	-0.8	0.44932896	1.175						
4	-0.6	0.54881164	1.175						
5	-0.4	0.67032005	1.175						
6	-0.2	0.81873075	1.175						
7	0	1	1.175						
8	0.2	1.22140276	1.175						
9	0.4	1.4918247	1.175						
10	0.6	1.8221188	1.175						
11	0.8	2.22554093	1.175						
12	1	2.71828183	1.175						
13									
14									
15									

Plot of f(x) = exp(x) and y = 1.175

The value of $\displaystyle\int_{-1}^{1} e^x \, dx$ is 2.35 and the area of the rectangle from $x = -1$ to 1 and having height 1.175

is 2.35. The two areas are the same.

7.4 Note the smoothing effect of the moving average as indicated by the dashed line. The solid line shows the original data.

	A	B	C	D	E	F	G	H	I
1	1	60							
2	2	70							
3	3	65							
4	4	55	62.5						
5	5	60	62.5						
6	6	60	60						
7	7	80	63.75						
8	8	85	71.25						
9	9	80	76.25						
10	10	80	81.25						
11	11	75	80						
12	12	75	77.5						
13	13	80	77.5						
14	14	85	78.75						
15	15	90	82.5						
16	16	85	85						
17	17	90	87.5						
18	18	100	91.25						
19	19	100	93.75						
20	20	100	97.5						

Plot of stock prices and moving average

7.5 The right Riemann sum approximation to the total revenue is shown in the following spreadsheet:

	C1			=	=SUM(B2:B11)*2
	A	B	C	D	
1	t	R(t)=2t+1	460		
2	2	5			
3	4	9			
4	6	13			
5	8	17			
6	10	21			
7	12	25			
8	14	29			
9	16	33			
10	18	37			
11	20	41			

The total revenue using the definite integral is $\int_{0}^{20}(2t+1)dt = 420$. For the full year, the total revenue

would be $\int_{0}^{365}(2t+1)dt = 133{,}590$.

7.6 The total value of the income stream is $\int_{0}^{5}R(t)dt = \int_{0}^{5}10{,}000dt = \$50{,}000$. The future value at the end

of 5 years is $e^{rb}\int_{a}^{b}R(t)e^{-rt}dt = e^{0.05(5)}\int_{0}^{5}10{,}000e^{-.05t}dt = \$56{,}805.08$. The Riemann sum approxima-

tion to the future value using 10 subdivisions is \$57,518.11 as shown below:

	B1			=	=EXP(0.25)*10000*EXP(-0.05*A1)	
	A	B	C	D	E	F
1	0	12840.25	57518.11			
2	0.5	12523.23				
3	1	12214.03				
4	1.5	11912.46				
5	2	11618.34				
6	2.5	11331.48				
7	3	11051.71				
8	3.5	10778.84				
9	4	10512.71				
10	4.5	10253.15				

If 100 subdivisions are used, the Riemann sum is \$56,876.12. If 1000 subdivisions are used, the Riemann sum is \$56,812.18.

7.7 The present value of this continuous income stream is $e^{ra}\int_a^b R(t)e^{-rt}\,dt = e^{0.05(0)}\int_0^5 10{,}000^{-0.05t}\,dt =$ $44,239.84. The midpoint Riemann sum approximation to the present value using 10 subdivisions is $44,238.69 as shown:

	B1		=	=10000*EXP(-0.05*A1)	
	A	B	C	D	E
1	0.25	9875.778	44238.69		
2	0.75	9631.944			
3	1.25	9394.131			
4	1.75	9162.189			
5	2.25	8935.973			
6	2.75	8715.343			
7	3.25	8500.161			
8	3.75	8290.291			
9	4.25	8085.603			
10	4.75	7885.969			

If 100 subdivisions are used, the Riemann sum is $44,239.83. If 1000 subdivision are used, the Riemann sum is $44,239.84. Note how close the midpoint Riemann sum approximation with 100 subdivisions is to the answer given by the definite integral for the present value.

7.8 The value of the integral is $\lim_{M\to\infty}\int_0^M e^{-x}\,dx = \lim_{M\to\infty}\left(1 - e^{-M}\right) = 1$. The graph of the integrand is as follows:

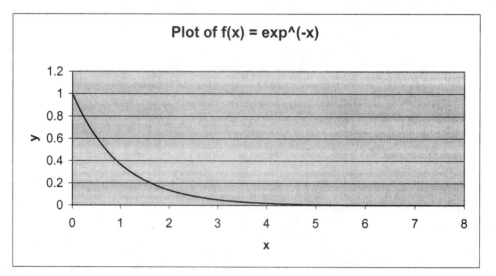

Plot of f(x) = exp^(-x)

The value of the left Riemann sum with $n = 100$ over the interval from $x = 0$ to $x = 10$ is 1.050785.

7.9 The integral is defined as $\displaystyle\lim_{r\to 1^+} \int_r^2 4(x-1)^{-1/2}\, dx = \lim_{r\to 1^+}\left[8 - 8(r-1)^{1/2}\right] = 8.$

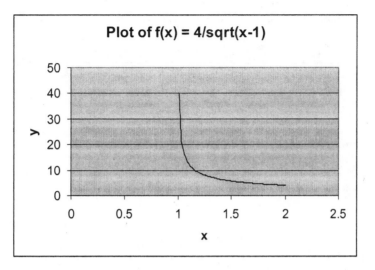

The value of the right Riemann sum with $n = 100$ is 7.435842. The value of the right Riemann sum with $n = 1{,}000$ is 7.817278.

7.10 The general solution is $y = \dfrac{1}{2}e^{x^2} - \cos x - \dfrac{1}{2}x^2 + 2x + C$.

The particular solution is $= y = \dfrac{1}{2}e^{x^2} - \cos x - \dfrac{1}{2}x^2 + 2x + \dfrac{1}{2}$.

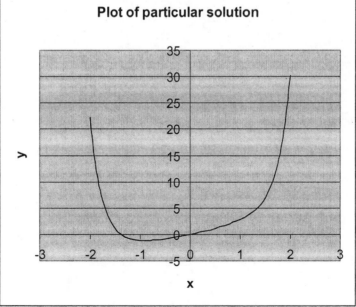

Functions of Several Variables

8.1 The following panel shows the setup for plotting the function. The values from cell B1 to F1 are x values and the values from cell A2 to A8 are y values. The expression in cell B2 shows how the equation is entered.

	B2		=	=B$1+$A2+B$1*$A2-3	
	A	B	C	D	E	F
1		0	0.5	1	1.5	2
2	0	-3	-2.5	-2	-1.5	-1
3	0.5	-2.5	-1.75	-1	-0.25	0.5
4	1	-2	-1	0	1	2
5	1.5	-1.5	-0.25	1	2.25	3.5
6	2	-1	0.5	2	3.5	5
7	2.5	-0.5	1.25	3	4.75	6.5
8	3	0	2	4	6	8

The plot is as follows:

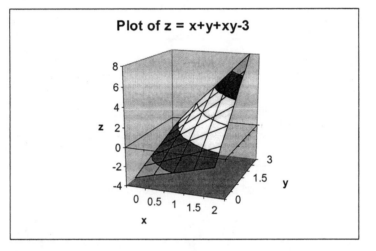

Plot of z = x+y+xy-3

8.2 The plot is as follows:

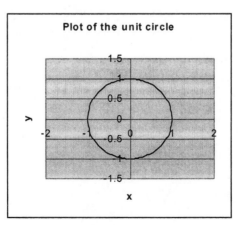

Plot of the unit circle

150

8.3 The equation of the top half of the sphere is $z = 5 + \sqrt{25 - x^2 - y^2}$. The construction of the top half of the sphere is accomplished as follows:

B2		= =5+SQRT(25-$A2^2-B$1^2)			
	A	B	C	D	E	F

	A	B	C	D	E	F
1		-1	-0.5	0	0.5	1
2	-1	9.795832	9.873397	9.898979	9.873397	9.795832
3	-0.5	9.873397	9.949747	9.974937	9.949747	9.873397
4	0	9.898979	9.974937	10	9.974937	9.898979
5	0.5	9.873397	9.949747	9.974937	9.949747	9.873397
6	1	9.795832	9.873397	9.898979	9.873397	9.795832

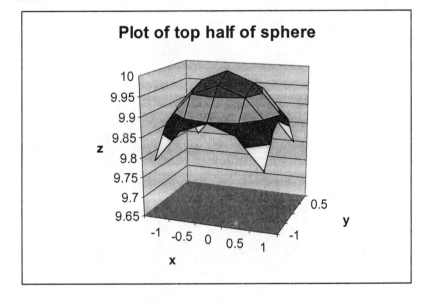

8.4 The first and second derivatives are

$$f_x = \frac{-x}{\sqrt{-x^2 - y^2 + 2y}}, f_{xx} = \frac{-x^2}{(-x^2 - y^2 + 2y)^{3/2}} - \frac{1}{\sqrt{-x^2 - y^2 + 2y}}, f_y = \frac{-2y + 2}{2\sqrt{-x^2 - y^2 + 2y}},$$

and $f_{yy} = \dfrac{-(-2y + 2)^2}{4(-x^2 - y^2 + 2y)^{3/2}} - \dfrac{1}{\sqrt{-x^2 - y^2 + 2y}}$.

The two first derivatives are equal to 0 when $x = 0$ and $y = 1$. The mixed partial derivative is equal to $f_{xy} = \dfrac{x(-2y + 2)}{2(-x^2 - y^2 + 2y)^{3/2}}$. The quantity, H, is equal to $H = f_{xx}(0,1)f_{yy}(0,1) - [f_{xy}(0,1)]^2 =$ $(-1)(-1) - (0) = 1$. Since H is positive and $f_{xx}(0,1) = -1$ is negative, by the second-derivative test, we know that f has a relative maximum at $(0, 1)$. That is, the point $(0, 1, 1)$ is a point in 3-dimensional space that corresponds to a relative maximum point located on the surface defined by $z = f(x, y) = \sqrt{-x - x^2 - y^2 + 2y}$.

B2		=	=SQRT(-(B$1^2)-($A2^2)+2*$A2)		
	A	B	C	D	E	F

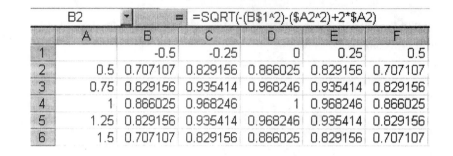

	A	B	C	D	E	F
1		-0.5	-0.25	0	0.25	0.5
2	0.5	0.707107	0.829156	0.866025	0.829156	0.707107
3	0.75	0.829156	0.935414	0.968246	0.935414	0.829156
4	1	0.866025	0.968246	1	0.968246	0.866025
5	1.25	0.829156	0.935414	0.968246	0.935414	0.829156
6	1.5	0.707107	0.829156	0.866025	0.829156	0.707107

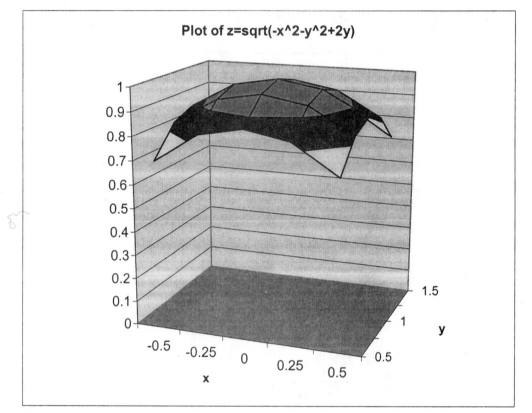

Plot of z=sqrt(-x^2-y^2+2y)

8.5 $\int_0^2 \int_0^2 x^2 + y^2 \, dxdy$ = 32/3 = 10.67. This is the volume under the surface, $z = x^2 + y^2$, for the region $0 < x < 2$ and $0 < y < 2$. For $\Delta x = \Delta y = 0.05$, the upper left and lower right parts of the spreadsheet are as follows:

B2		=	=$A2^2+B$1^2
	A	B	C	D
1		0	0.05	0.1
2	0	0	0.0025	0.01
3	0.05	0.0025	0.005	0.0125
4	0.1	0.01	0.0125	0.02
5	0.15	0.0225	0.025	0.0325
6	0.2	0.04	0.0425	0.05

AM	AN	AO	AP
5.9825	6.17	6.3625	
6.145	6.3325	6.525	
6.3125	6.5	6.6925	
6.485	6.6725	6.865	
6.6625	6.85	7.0425	
6.845	7.0325	7.225	
7.0325	7.22	7.4125	
7.225	7.4125	7.605	
188.25	195.75	203.45	4108

The double Riemann sum for $\Delta x = \Delta y = 0.05$ is $4108(0.05)(0.05) = 10.27$. For $\Delta x = \Delta y = 0.01$, the double Riemann sum value is $105868(0.01)(0.01) \approx 10.59$.

8.6 The upper left and lower right portion of the worksheet for $\Delta x = \Delta y = 0.1$ are as follows:

B2		=	=EXP(-(B$1^2)-($A2^3))
	A	B	C	D	E
1		0	0.1	0.2	0.3
2	0	1	0.99005	0.960789	0.913931
3	0.1	0.999	0.98906	0.959829	0.913018
4	0.2	0.992032	0.982161	0.953134	0.906649
5	0.3	0.973361	0.963676	0.935195	0.889585
6	0.4	0.938005	0.928672	0.901225	0.857272

R	S	T	U	V
0.008591	0.006177	0.004353	0.003006	
0.004972	0.003574	0.002519	0.00174	
0.002645	0.001902	0.00134	0.000926	
0.001286	0.000925	0.000652	0.00045	
0.000568	0.000409	0.000288	0.000199	
0.000227	0.000163	0.000115	7.93E-05	
8.12E-05	5.84E-05	4.11E-05	2.84E-05	
0.728932	0.524046	0.369289	0.255081	87.79689

The double Riemann sum for $\Delta x = \Delta y = 0.1$ is $87.79689(0.1)(0.1) \approx 0.878$. For $\Delta x = \Delta y = 0.05$, the double Riemann sum value is $332.8863(0.05)(0.05) \approx 0.832$.

Index

S

SIN function, 25
Slope of a line, 7-8
    Finding graphically, 8
Solver Add-in Tool, 117
Sphere
    Plotting in three dimensions,112-113
SQRT (square root) function, 3, 13, 48, 101,
    105, 109, 112
Subscripts, 2
Surface of an elliptic paraboloid, 109-112

T

TAN function, 27
Text concatenation, 3
Total revenue of a continuous income stream,
    95-96
Total social gain, 91
Trendline, 13

V

Velocity and acceleration, 57, 72
Vertex, 17-18

W

Worksheet function, 12

X

XY (scatter) plot, 5-6, 10

Z

Zooming in on stationary points, 51-53